A Practical Introduction to Finite Element Analysis

A Practical Introduction to Finite Element Analysis

Y K CHEUNG
Ph.D., D.Sc., F.I.C.E., F.I.Struct.E., F.A.S.C.E.
Professor of Civil Engineering
University of Hong Kong

and

M F YEO
M.Sc., Ph.D., M.I.E.Aust.
Senior Lecturer in Civil Engineering
University of Adelaide
Australia

PITMAN

PITMAN PUBLISHING LIMITED
39 Parker Street, London WC2B 5PB

North American Editorial Office
1020 Plain Street, Marshfield, Massachusetts

Associated Companies
Fearon Pitman Publishers Inc, San Francisco
Copp Clark Pitman, Toronto
Pitman Publishing New Zealand Ltd, Wellington
Pitman Publishing Pty Ltd, Melbourne

© Y K Cheung & M F Yeo 1979

LIBRARY OF CONGRESS CATALOGING IN PUBLICATION DATA
Cheung, Y.K.
A practical introduction to finite element analysis
Includes bibliographical references and index
1. Finite element method   2. Engineering data processing
I. Yeo, M.F., joint author   II. Title
TA347.F5C47   624.171   79-4262

ISBN 0 273 01083 2 hardcover
ISBN 0 273 01080 8 flexicover

Reproduced and printed by photolithography
in Great Britain at Biddles of Guildford

Preface

In this book the authors have attempted to present an introduction to
the finite element method which has become the most powerful and
versatile tool available to engineers. Unlike a number of other texts
which dwell at length on the theoretical development and applications
of the finite element method, the emphasis of this text is on various
practical aspects of the method such as (i) the detailed description of
various techniques used in the choice of displacement functions, stiffness formulation, numerical integration, assembly, solution and mesh
generation; (ii) the detailed explanation of computer programs at
various levels of development and comments for nearly all program
statements; and (iii) intermediate output at all stages of program
execution and final output through several case studies so that the
readers can see what is happening at every step. The book is intended
for final-year students, post-graduate students and for practising
engineers. The material included is largely self-contained and the
readers are required only to have some basic knowledge of matrix
algebra and Fortran programming. To keep the book down to a reasonable
size, only plane stress problems are discussed here, although obviously
the basic numerical and programming techniques are also applicable to
three-dimensional elasticity, to ideal flow and to other problems.
  The theory of the finite element method is considered in Chapter 1,
and considerable effort is devoted to the topic of displacement
functions. In Chapter 2 a plane stress triangular element is derived
and a simple but workable computer program is described in a step-by-
step fashion. The isoparametric concept of formulating elements is
introduced in Chapter 3 and numerical integration schemes and different
techniques used in reducint the amount of efforts in calculating an
element stiffness matrix are discussed. In Chapter 4, the front
solution technique, which can be considered as one of the most (if not
the most) effective solution schemes, is presented in great detail and
simple illustrative examples, with all computation steps shown, are
used in helping the readers to understand the manipulations. Chapter 5
gives the complete listing for an advance computer program and the
sample input and output for an illustrative problem. Finally, two
simple mesh generation techniques are introduced in Chapter 6 and their
corresponding routines are also listed.

We are indebted to Mr G. Sved of the University of Adelaide for reading the manuscript and for making valuable suggestions for improving the presentation. Thanks are due to Mr S. Swaddiwudhipong and Mr C. Rigon, both doctoral students at the University of Adelaide, for checking parts of the manuscript, and to Mrs G. Stock, for her very skilful typing of the manuscript.

Finally we are deeply grateful to Mrs Yuk Bau Cheung for her constant support and encouragement in this effort.

August 1978

Y K Cheung
M F Yeo

Contents

1 The Finite Element Method

1.1 Introduction

It is a well-known fact that the finite element method, which by now has become the most powerful tool of analysis and is applicable to a wide range of problems, started as an extension of the stiffness or displacement method, in which a skeletal structure is assumed to be made up of an assemblage of one-dimensional elements such as bar elements (axial actions), beam elements (bending actions) and frame elements (axial, bending and torsional actions).

In the stiffness method for skeletal structures the elements of an actual structure are connected together at discrete joints, and equations of equilibrium involving external loads and member end forces are established at all joints to solve for joint displacements. The relationship between the end forces and end displacements of each member is represented by the stiffness matrix, which can be derived directly through the various energy theorems such as the principle of virtual work or principle of minimum total potential energy.

A similar situation exists for two- and three-dimensional elements for solid mechanics in which elements of some geometrical shapes such as triangles, quadrilaterals and hexahedra are connected at some artificially created points, and stiffness relationships between nodal displacements and equivalent nodal forces are derived in the same manner. In this way a continuum with infinite degrees of freedom can be discretized into an equivalent system with finite degrees of freedom.

In recent years the finite element method has been applied widely to non-structural problems, and the formulation is now mainly based on variational principles or weighted residual procedures. A detailed presentation can be found in a text by Zienkiewicz [1].

Throughout this book energy principles will be used for the derivation of the stiffness and load matrices. All elements and example problems presented later are in the field of two-dimensional elasticity, so as to keep the presentation as simple as possible. However, much of the material discussed herein is also directly applicable to three-dimensional elasticity and even to flow problems.

1.2 Principle of virtual work

The principle relates to two distinct and separate systems in which the first is a set of forces in equilibrium (with P and σ as the external forces and internal stresses respectively) and the second a set of geometrically compatible deformations (with Δ and ε as the displacements and strains, respectively). The principle states that for any system in equilibrium, the external virtual work must be equal to the internal virtual work, i.e.

System in equilibrium

$$\sum P.\Delta = \int_V \sigma.\varepsilon \, dv \tag{1.1}$$

Geometrically compatible deformations

Note that, in practice, one of the systems always relates to a real or actual structure in which some sort of solution is required, while the other is an imaginary or virtual system. Therefore, it is possible to have the option of establishing:

(i) Theorem of virtual forces in which a real system of displacements and strains is coupled to a virtual system of forces and stresses. Applying Eqn.(1.1), we have

$$\sum (\text{Virtual external forces}) \, . \, (\text{Actual displacements})$$
$$= \int_V (\text{Virtual stresses}) \, . \, (\text{Actual strains}) \, dv \tag{1.2}$$

(ii) Theorem of virtual displacements in which a real system of forces and stresses is coupled to a virtual system of displacements and strains. Again, using Eqn.(1.1).

$$\sum (\text{Real external forces}) \, . \, (\text{Virtual displacements})$$
$$= \int_V (\text{Real stresses}) \, . \, (\text{Virtual strains}) \, dv \tag{1.3}$$

Eqn.(1.2) is used to calculate displacements and leads to the unit load theorem while Eqn.(1.3) is used to calculate external forces and leads to the unit displacement theorem.

1.3 Principle of minimum total potential energy

The total potential energy of a system is defined as

$$\phi = U + W \tag{1.4}$$

in which W is the potential energy of the external forces in the deformed configuration and is defined as

$$W = -\sum P.\Delta \tag{1.5}$$

and U is the strain energy of the deformed structure and is given by

$$U = \int_V \left(\int \sigma d\varepsilon \right) dv \tag{1.6}$$

If the deformed system is now taken as the real or actual system, and a set of small geometrically compatible displacements $\delta\Delta$ as the virtual system, then, by virtue of Eqn.(1.3), it is possible to establish that

$$\sum P.\delta\Delta = \int_V (\sigma.\delta\varepsilon)\,dv \qquad (1.7)$$

However, due to the imposed virtual displacements, the actual system will also undergo a change in total potential energy of

$$\delta\phi = \delta U + \delta W$$

$$= \int_V (\sigma.\delta\varepsilon)\,dv - \sum P.\delta\Delta \qquad (1.8)$$

By comparing Eqn.(1.7) with Eqn.(1.8), it can be concluded that

$$\delta\phi = 0. \qquad (1.9)$$

In other words, the total potential energy of a system in equilibrium is stationary, and, because it can be further proved that the energy is always a minimum for stable structures, the statement is called the principle of stationary total potential energy or, more often, principle of minimum total potential energy. The principle can also be alternatively expressed as 'of all compatible displacements satisfying given boundary conditions, those which satisfy the equilibrium conditions make the total potential energy assume a stationary value'. A solution satisfying both equilibrium and compatibility is of course the correct solution for a linearly elastic problem. However, such solutions are difficult, if not impossible, to work out for the majority of cases and researchers have to resort to approximate solutions by assuming compatible displacements with undetermined parameters and determining the values of the parameters such that the total potential energy of the system is a minimum.

It follows that if a set of trial displacement functions with unknown parameters Δ_i is used to approximate the actual displacements of a system, then it is possible to determine such Δ_i by minimizing the total potential energy, i.e. by performing the operations

$$\frac{\partial\phi}{\partial\Delta_i} = 0 \qquad (1.10)$$

Eqn.(1.10) has been used extensively in the derivation of finite element stiffness matrices.

1.4 Stiffness matrix of a bar member

Bar members are primarily used in trusses, and, by definition, they are assumed to be acted upon only by axial forces. The stiffness coefficients of a bar member can be derived directly from Hooke's Law, and referring to Fig.1.1(a) it is seen that if end 2 is fixed and end 1 is allowed to move, then

$$P_{x1} = \frac{EA}{\ell}u_1$$

$$P_{x2} = -\frac{EA}{\ell}u_1 \qquad (1.11)$$

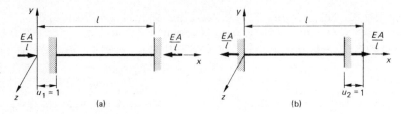

Fig.1.1 A bar member

Similarly, it can be concluded from Fig.1.1(b) that if end 1 is now
fixed but end 2 allowed to move,

$$P_{x1} = -\frac{EA}{\ell} u_2$$

$$P_{x2} = \frac{EA}{\ell} u_2$$

(1.12)

The complete stiffness relationship is now obtained by combining Eqn.
(1.11) and Eqn.(1.12), and is given as

$$\begin{Bmatrix} P_{x1} \\ P_{x2} \end{Bmatrix} = \frac{EA}{\ell} \begin{bmatrix} 1 & -1 \\ -1 & 1 \end{bmatrix} \begin{Bmatrix} u_1 \\ u_2 \end{Bmatrix} \qquad (u_1 - u_2) \qquad -u_1 + u_2$$

(1.13)

1.5 Stiffness matrix of a beam member

A beam member is concerned with bending action only and its stiffness
relationship is just a simple extension of the well-known slope-
deflection equations, which are of the form

$$M_1 = (6EI/\ell^2)v_1 + (4EI/\ell)\theta_1 - (6EI/\ell^2)v_2 + (2EI/\ell)\theta_2$$

$$M_2 = (6EI/\ell^2)v_1 + (2EI/\ell)\theta_1 - (6EI/\ell^2)v_2 + (4EI/\ell)\theta_2$$

(1.14(a))

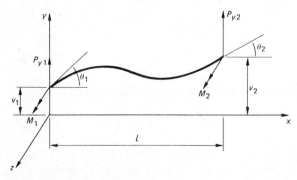

Fig.1.2 A beam member

The end shear forces P_{y1} and P_{y2} (Fig.1.2), which are equal and opposite,
form a couple equal to the sum of the end moments. Therefore, from
Eqn.(1.14(a)), the shear forces are

$$P_{y1} = -P_{y2} = (M_1 + M_2)/\ell$$

$$= (12EI/\ell^3)v_1 + (6EI/\ell^2)\theta_1 - (12EI/\ell^3)v_2 + (6EI/\ell^2)\theta_2$$

(1.14(b))

Writing Eqn.(1.14) in matrix form, the stiffness relationships for a beam become

$$
\begin{Bmatrix} P_{y1} \\ M_1 \\ P_{y2} \\ M_2 \end{Bmatrix} = \begin{bmatrix} 12EI/\ell^3 & 6EI/\ell^2 & -12EI/\ell^3 & 6EI/\ell^2 \\ 6EI/\ell^2 & 4EI/\ell & -6EI/\ell^2 & 2EI/\ell \\ -12EI/\ell^3 & -6EI/\ell^2 & 12EI/\ell^3 & -6EI/\ell^2 \\ 6EI/\ell^2 & 2EI/\ell & -6EI/\ell^2 & 4EI/\ell \end{bmatrix} \begin{Bmatrix} v_1 \\ \theta_1 \\ v_2 \\ \theta_2 \end{Bmatrix}
\tag{1.15}
$$

Note that the right-handed coordinate system is used here and will be used throughout the rest of the book.

1.6 Finite element procedure

Before we go any further into detailed derivations, it is perhaps worthwhile to point out that there are three types of commonly used elements. The displacement element using assumed displacement patterns form the overwhelming majority of all the known finite elements, while the equilibrium elements [2] (based on assumed stress patterns) and the hybrid element [3] (based on both assumed displacements and stresses) are used to a much lesser extent. In this book, only displacement elements will be discussed, and, since the assumed displacement patterns over the whole area of an element are described by the so-called displacement functions, the reader will note later that a great deal of attention is paid to the choice of suitable displacement functions for various finite elements.

As mentioned previously, the finite element method started as an extension of the stiffness method and was applied to two- and three-dimensional problems in structural mechanics. However, unlike skeletal structures, there are no well-defined joints where equilibrium of forces can be established and, therefore, the continuum must be discretized into a number of elements of arbitrary shapes and also artificial joints or nodes must be created. In this way the continuum is approximated by a system with a finite degree of freedom and a numerical solution can be achieved. A detailed description of the finite element procedure will be given below.

(i)   The continuum is divided into two- or three-dimensional finite elements, which are separated by straight or curved lines (two-dimensional) or by flat or curved surfaces (three-dimensional).

Since each element is allowed to have its own shape, size, material properties and thickness, it is not surprising that the method is admirably suitable for problems involving non-homogeneous properties, irregular geometry, complex support conditions, and combination of various types of loadings.

In many cases only one type of element is used for one problem, but it is also possible to mix elements of different types such as a beam element connected to a plate bending element because of physical requirements, or higher-order elements connected to lower-order elements in stress concentration problems.

(ii)   The elements are assumed to be inter-connected at a discrete
       number of nodal points situated at element corners or on element
       boundaries, although occasionally internal nodes (Fig.1.3(a))
       which are not connected to any other element may also be present.

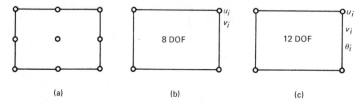

(a)   (b)   (c)

Fig.1.3 Rectangular plane stress elements

       The nodal degrees of freedom (which can vary from node to node
       within each element), called nodal-displacement parameters,
       normally refer to the displacements and their first partial
       derivatives (rotations) at a node, but very often may include
       other terms such as stresses, strains and second, or even
       higher, partial derivatives. As an example, the rectangular
       plane stress element of Fig.1.3(b) (bilinear element) has eight
       degrees of freedom representing the translations u and v at the
       four corner nodes in the x and y directions, while the higher-
       order plane stress element (beam type element) of Fig.1.3(c)
       has four additional rotational nodal degrees of freedom.
(iii) A displacement function in terms of the coordinate variables x,
       y and the nodal displacement parameters (e.g. u_i, v_i in Fig.1.3)
       is chosen to represent the displacement variations within each
       element, and, by using the principle of virtual work or the
       principle of minimum total potential energy discussed previously,
       a stiffness matrix relating the nodal 'forces' to the nodal
       'displacements' can be derived. Such a displacement function
       will try to approximate the actual displacement field over the
       whole element.

1.7 Displacement functions

From the above discussions, it is evident that the choice of suitable
displacement functions is the most important part of the whole procedure.
A good displacement function will lead to an element of high accuracy
and with converging characteristics, and conversely a wrongly chosen
displacement function yielding poor or non-converging results, and at
times even worse, solutions which converge to an incorrect answer. The
latter phenomenon was observed and discussed by Clough [4] but fortunately it is not a common occurrence.
     A displacement function is either given as (i) a simple polynomial
with undetermined coefficients which are subsequently transformed to
become the relevant nodal displacement parameters, or (ii) directly in
terms of shape functions, which are usually osculatory polynomials [5]
having zero values at all other nodes of an element but unit values for
the displacement or its partial derivatives at the node under consideration. Physically, the shape function associated with a nodal displacement parameter gives us the displacement field over an element when that

particular nodal displacement parameter is given a unit value and all other nodal displacement parameters are given zero values. Some examples of shape functions are given in Fig.1.4. Thus a displacement function can either be given (from (i) above) as

$$f = A_1 + A_2 x + A_3 y + \ldots \tag{1.16(a)}$$

in which A_1, A_2 etc. are undetermined polynomial constants, or from (ii) as

$$f = N_1(x,y)f_1 + N_2(x,y)f_2 + \ldots \tag{1.16(b)}$$

in which f_1, f_2 etc. are the nodal displacement parameters and N_1, N_2 etc. the corresponding shape functions.

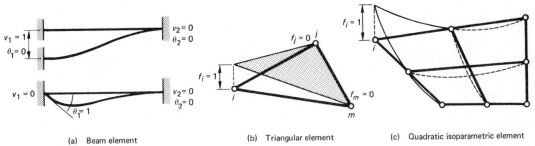

(a) Beam element   (b) Triangular element   (c) Quadratic isoparametric element

Fig.1.4 Examples of shape functions

At this stage it is appropriate to become familiar with the following concepts which will be helpful to us in the construction of displacement functions.

1.7.1 THE PASCAL TRIANGLE

For displacement functions of two-dimensional elements given in terms of simple polynomials, the Pascal triangle (Fig.1.5) is a useful aid for determining the combination of terms which should be used. The terminology used here is somewhat different from the one used in mathematics which deals with the coefficients of the binomial theorem.

$$1$$
$$x \qquad y$$
$$x^2 \qquad xy \qquad y^2$$
$$x^3 \qquad x^2 y \qquad xy^2 \qquad y^3$$
$$x^4 \qquad x^3 y \qquad x^2 y^2 \qquad xy^3 \qquad y^4$$
$$x^5 \qquad - \qquad - \qquad - \qquad - \qquad y^5$$

Fig.1.5 Pascal Triangle

In Table 1.1 some examples on the use of the Pascal triangle are given. It should be noticed that for general purpose elements symmetry in the x, y terms is always preserved and that normally all the terms for a lower degree polynomial should be used before taking up more terms from a higher degree polynomial.

TABLE 1.1 *Typical finite elements with their degrees of freedom (DOF) and their displacement functions*

Element type	Pascal triangle	Remarks
Constant strain triangle u – 3 DOF v – 3 DOF	1 $x \quad y$ $x^2 \quad xy \quad y^2$	Linear variation of u and v
Plane stress rectangle u – 4 DOF v – 4 DOF	1 $x \quad y$ $x^2 \quad xy \quad y^2$	Linear variation of u and v along edges of element required. xy is the appropriate fourth term since it degenerates either to x or y only along the edges.
Bending rectangle w – 12 DOF	1 $x \quad y$ $x^2 \quad xy \quad y^2$ $x^3 \quad x^2y \quad xy^2 \quad y^3$ $x^4 \quad x^3y \quad x^2y^2 \quad xy^3 \quad y^4$ $x^5 \quad x^4y \quad x^3y^2 \quad x^2y^3 \quad xy^4 \quad y^5$	Cubic variation along edges of element required. x^3y and xy^3 degenerate to cubic functions of x and y only along the edges.
Rectangular beam type plane stress element u – 4 DOF v – 8 DOF	1 $x \quad y$ $x^2 \quad xy \quad y^2$ 1 $x \quad y$ $x^2 \quad xy \quad y^2$ $x^3 \quad x^2y \quad xy^2 \quad y^3$ $x^4 \quad x^3y \quad x^2y^2 \quad xy^3 \quad y^4$	Linear variation of u along edges. Variation of v linear in y and cubic in x so as to simulate correctly beam behaviour.
Rectangular plane stress element with corner and midside nodes u – 8 DOF v – 8 DOF	1 $x \quad y$ $x^2 \quad xy \quad y^2$ $x^3 \quad x^2y \quad xy^2 \quad y^3$	Parabolic variation of u and v along edges. x^3 and y^3 excluded.

1.7.2 LAGRANGE POLYNOMIAL

Lagrange polynomials (collocation polynomials) [5] are often used for the construction of shape functions of elements in which only function values but not derivatives are specified at the nodes. The basic form of the Lagrange polynomial in a single coordinate system with n nodes is

$$f(x) = \sum_{i=0}^{n} \mathcal{L}_i^n(x) f_i \tag{1.17}$$

where $\mathcal{L}_i^n(x)$ is called the Lagrange multiplier function and is given by

$$\mathcal{L}_i^n(x) = \frac{(x - x_0)(x - x_1) \ldots (x - x_{i-1})(x - x_{i+1}) \ldots (x - x_n)}{(x_i - x_0)(x_i - x_1) \ldots (x_i - x_{i-1})(x_i - x_{i+1}) \ldots (x_i - x_n)} \tag{1.18}$$

It is obvious that $\mathcal{L}_i^n(x)$ possesses the properties of

$$\mathcal{L}_i^n(x_k) \begin{cases} = 0 & k \neq i \\ = 1 & k = i \end{cases}$$

and thus fits in with the definition of a shape function. Some examples of Lagrange polynomials are given in Table 1.2.

It is also possible to apply the Lagrange polynomials to shape functions involving two or even three coordinates. Thus the shape functions for a two-dimensional problem would be

$$f(x,y) = \sum_{i=0}^{n} \sum_{j=0}^{m} \mathcal{L}_i^n(x) \mathcal{L}_j^m(y) f_{ij} \tag{1.19}$$

where n and m stand for the number of subdivisions or arguments in the x and y directions respectively.

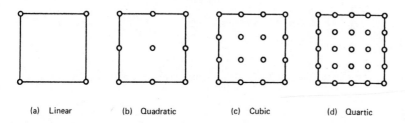

(a) Linear   (b) Quadratic   (c) Cubic   (d) Quartic

Fig.1.6 Family of Lagrange elements

The shape functions given in Eqn.(1.19) in fact form the basis of a family of plane stress elements given in Fig.1.6. However, the usefulness of this family of elements is severely limited by the presence of a large number of internal nodes, and in fact most research workers have restricted their attention to the linear element only.

One Dimensional

TABLE 1.2 *Lagrange polynomials (single coordinate only)*

Linear

$$\mathcal{L}_0^1(x) = \frac{x - x_1}{0 - x_1} = 1 - \frac{x}{\ell}$$

$$\mathcal{L}_1^1(x) = \frac{x}{x_1} = \frac{x}{\ell}$$

Parabolic

$$\mathcal{L}_0^2(x) = \frac{(x - x_1)(x - x_2)}{(0 - x_1)(0 - x_2)} = \frac{(x - x_1)(x - x_2)}{x_1 x_2}$$

$$= \left(2\frac{x}{\ell} - 1\right)\left(\frac{x}{\ell} - 1\right) = \left(1 - 2\mathcal{L}_1^1\right)\mathcal{L}_0^1$$

$$\mathcal{L}_1^2(x) = \frac{(x - 0)(x - x_2)}{(x_1 - 0)(x_1 - x_2)} = \frac{x(x - x_2)}{x_1(x_1 - x_2)}$$

$$= 4\frac{x}{\ell}\left(1 - \frac{x}{\ell}\right) = 4\mathcal{L}_0^1\mathcal{L}_1^1$$

$$\mathcal{L}_2^2(x) = \frac{(x - 0)(x - x_1)}{(x_2 - 0)(x_2 - x_1)} = \frac{x(x - x_1)}{x_2(x_2 - x_1)}$$

$$= \frac{x}{\ell}\left(2\frac{x}{\ell} - 1\right) = \mathcal{L}_1^1\left(2\mathcal{L}_1^1 - 1\right)$$

Cubic

$$\mathcal{L}_0^3(x) = \frac{(x - x_1)(x - x_2)(x - x_3)}{(0 - x_1)(0 - x_2)(0 - x_3)}$$

$$= \frac{(x_1 - x)(x_2 - x)(x_3 - x)}{x_1 x_2 x_3}$$

$$= \frac{1}{2}\left(1 - \frac{3x}{\ell}\right)\left(2 - \frac{3x}{\ell}\right)\left(1 - \frac{x}{\ell}\right)$$

$$= \frac{1}{2}\left(1 - 3\mathcal{L}_1^1\right)\left(2 - 3\mathcal{L}_1^1\right)\mathcal{L}_0^1$$

$$\mathcal{L}_1^3(x) = \frac{(x - 0)(x - x_2)(x - x_3)}{(x_1 - 0)(x_1 - x_2)(x_1 - x_3)} = \frac{x(x - x_2)(x - x_3)}{x_1(x_1 - x_2)(x_1 - x_3)}$$

$$= \frac{9}{2}\frac{x}{\ell}\left(3\frac{x}{\ell} - 2\right)\left(\frac{x}{\ell} - 1\right) = \frac{9}{2}\mathcal{L}_1^1\left(2 - 3\mathcal{L}_1^1\right)\mathcal{L}_0^1$$

$$\mathcal{L}_2^3(x) = \frac{(x - 0)(x - x_1)(x - x_3)}{(x_2 - 0)(x_2 - x_1)(x_2 - x_3)} = \frac{x(x - x_1)(x - x_3)}{x_2(x_2 - x_1)(x_2 - x_3)}$$

$$= \frac{9}{2}\frac{x}{\ell}\left(3\frac{x}{\ell} - 1\right)\left(1 - \frac{x}{\ell}\right) = \frac{9}{2}\mathcal{L}_1^1\left(3\mathcal{L}_1^1 - 1\right)\mathcal{L}_0^1$$

$$\mathcal{L}_3^3(x) = \frac{(x - 0)(x - x_1)(x - x_2)}{(x_3 - 0)(x_3 - x_1)(x_3 - x_2)} = \frac{x(x - x_1)(x - x_2)}{x_3(x_3 - x_1)(x_3 - x_2)}$$

$$= \frac{1}{2}\frac{x}{\ell}\left(3\frac{x}{\ell} - 1\right)\left(3\frac{x}{\ell} - 2\right) = \frac{1}{2}\mathcal{L}_1^1\left(3\mathcal{L}_1^1 - 1\right)\left(3\mathcal{L}_1^1 - 2\right)$$

1.7.3 HERMITIAN POLYNOMIAL

The Hermitian or osculatory polynomials [5] not only agree in value with a given function at specified locations, as is the case for collocation polynomials, but their derivatives will also match with the derivatives of the given functions up to any given order at the said locations. With only a first-order osculation, the Hermitian polynomial is the simplest one and has the form

$$f(x) = \sum_{i=0}^{n} U_i(x) f_i + \sum_{i=0}^{n} V_i(x) f_i'$$ (1.20)

where f_i and f_i' are the function value and the first derivative respectively at point i, and

$$U_i(x) = \left[1 - 2\frac{d}{dx}\left(\mathscr{L}_i^n(x)\right)_{x=x_i}(x - x_i)\right]\left[\mathscr{L}_i^n(x)\right]^2$$

$$V_i(x) = (x - x_i)\left[\mathscr{L}_i^n(x)\right]^2$$ (1.21)

The most commonly used Hermitian polynomial is the so-called beam function involving two points only, and taking the distance between the two points as ℓ, then we have

$$\mathscr{L}_0^n(x) = \mathscr{L}_0^1(x), \quad \mathscr{L}_1^n(x) = \mathscr{L}_1^1(x)$$

(*see* Table 1.2). From Eqn.(1.21),

$$U_0(x) = \left(1 + \frac{2x}{x_1}\right)\left(\frac{x^2 - 2xx_1 + x_1^2}{x_1^2}\right)$$

$$= 1 - \frac{3x^2}{x_1^2} + \frac{2x^3}{x_1^3}$$

$$= 1 - \frac{3x^2}{\ell^2} + \frac{2x^3}{\ell^3}$$

$$U_1(x) = \left(1 - \frac{2x}{x_1} + 2\right)\left(\frac{x}{x_1}\right)^2 = \frac{3x^2}{x_1^2} - \frac{2x^3}{x_1^3} = \frac{3x^2}{\ell^2} - \frac{2x^3}{\ell^3}$$ (1.22(a))

Similarly,

$$V_0(x) = x\left(1 - 2\frac{x}{\ell} + \frac{x^2}{\ell^2}\right)$$ (1.22(b))

$$V_1(x) = x\left(\frac{x^2}{\ell^2} - \frac{x}{\ell}\right)$$

Another commonly used Hermitian polynomial is the one involving three equally spaced points with $x_0 = 0$, $x_1 = \ell/2$ and $x_2 = \ell$, and, in this case, $\mathcal{L}_0^2(x)$, $\mathcal{L}_1^2(x)$ and $\mathcal{L}_2^2(x)$ (*see* Table 1.2) will be the appropriate terms for $\mathcal{L}_i^n(x)$ in Eqn.(1.21).

$$U_0(x) = \left(1 + 6\frac{x}{\ell}\right)\left(4\frac{x^4}{\ell^4} - 12\frac{x^3}{\ell^3} + 13\frac{x^2}{\ell^2} - 6\frac{x}{\ell} + 1\right)$$

$$= 24\frac{x^5}{\ell^5} - 68\frac{x^4}{\ell^4} + 66\frac{x^3}{\ell^3} - 23\frac{x^2}{\ell^2} + 1$$

$$V_0(x) = x\left(4\frac{x^4}{\ell^4} - 12\frac{x^3}{\ell^3} + 13\frac{x^2}{\ell^2} - 6\frac{x}{\ell} + 1\right) \tag{1.23(a)}$$

Similarly, by taking $i = 1$ and 2 in Eqn.(1.21), it is possible to arrive at

$$U_1(x) = 16\frac{x^4}{\ell^4} - 32\frac{x^3}{\ell^3} + 16\frac{x^2}{\ell^2}$$

$$V_1(x) = x\left(16\frac{x^4}{\ell^4} - 40\frac{x^3}{\ell^3} + 32\frac{x^2}{\ell^2} - 8\frac{x}{\ell}\right)$$

$$\tag{1.23(b)}$$

$$U_2(x) = -24\frac{x^5}{\ell^5} + 52\frac{x^4}{\ell^4} - 34\frac{x^3}{\ell^3} + 7\frac{x^2}{\ell^2}$$

$$V_2(x) = x\left(4\frac{x^4}{\ell^4} - 8\frac{x^3}{\ell^3} + 5\frac{x^2}{\ell^2} - \frac{x}{\ell}\right)$$

The above functions have been used in bending elements (finite strips) with mid-side nodal lines [6]. The use of such higher-order functions tends to improve the accuracy of the finite elements.

1.7.4 AREA COORDINATES

All the previous discussions are concerned with Cartesian (x,y) coordinates, and the examples given on two-dimensional problems are valid only for rectangular elements. For general quadrilateral elements with straight or curved sides similar functions can be used, but they should be in curvilinear (ξ,η) coordinates, and a detailed discussion will be given later in Chapter 3. The Cartesian coordinates are also not very convenient for triangular elements, and a special type of coordinate system called area coordinates should be used.

Referring to Fig.1.7 it is seen that the internal point ℓ will divide the triangle ijm into three smaller triangles, and depending on the position of the point ℓ, the area of each one of the triangles ℓim, ℓmj and ℓji can vary from zero to Δ, which is the area of the triangle ijm. In other words, the ratios A_i/Δ, A_j/Δ and A_m/Δ will take up any value between zero and unity in the same way as a first-order Lagrange polynomial. These ratios are called area coordinates, and they are defined by

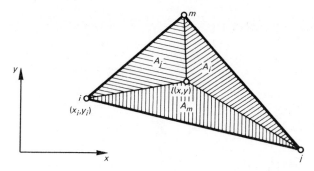

Fig.1.7 A triangle and the area coordinate system

$$\left. \begin{array}{l} \mathcal{L}_i = A_i/\Delta = (a_i + b_i x + c_i y)/2\Delta \\ \mathcal{L}_j = A_j/\Delta = (a_j + b_j x + c_j y)/2\Delta \\ \mathcal{L}_m = A_m/\Delta = (a_m + b_m x + c_m y)/2\Delta \end{array} \right\} \qquad (1.24)$$

in which

$$\left. \begin{array}{l} a_i = x_j y_m - x_m y_j \\ b_i = y_j - y_m \\ c_i = x_m - x_j \end{array} \right\} \qquad (1.25)$$

$$2\Delta = \det \begin{bmatrix} 1 & x_i & y_i \\ 1 & x_j & y_j \\ 1 & x_m & y_m \end{bmatrix} = 2(\text{area of triangle } ijm),$$

x_i, y_i, etc. are the nodal coordinates and a_j, b_j, c_j, etc. can be computed through a cyclic permutation of the subscripts.

It is interesting to note that some other useful information can be derived from Eqn.(1.24), which in matrix form is

$$\begin{Bmatrix} \mathcal{L}_i \\ \mathcal{L}_j \\ \mathcal{L}_m \end{Bmatrix} = \frac{1}{2\Delta} \begin{bmatrix} a_i & b_i & c_i \\ a_j & b_j & c_j \\ a_m & b_m & c_m \end{bmatrix} \begin{Bmatrix} 1 \\ x \\ y \end{Bmatrix} \qquad (1.26)$$

Solving for 1, x, y,

$$\begin{Bmatrix} 1 \\ x \\ y \end{Bmatrix} = \begin{bmatrix} 1 & 1 & 1 \\ x_i & x_j & x_m \\ y_i & y_j & y_m \end{bmatrix} \begin{Bmatrix} \mathcal{L}_i \\ \mathcal{L}_j \\ \mathcal{L}_m \end{Bmatrix} \qquad (1.27(a))$$

TABLE 1.3 *Shape functions for triangular elements*

	Lagrange polynomial (Table 1.2)	Triangular elements in terms of area coordinates

Linear
(Fig.1.8(a))

$$\mathcal{L}_0^1(x)$$

$$\mathcal{L}_1^1(x)$$

$N_1 = \mathcal{L}_1$

$N_2 = \mathcal{L}_2$

$N_3 = \mathcal{L}_3$

Quadratic
(Fig.1.8(b))

$$\mathcal{L}_0^2(x) = \left(1 - 2\mathcal{L}_1^1\right)\mathcal{L}_0^1$$

$$\mathcal{L}_1^2(x) = 4\mathcal{L}_0^1\mathcal{L}_1^1$$

$$\mathcal{L}_2^2(x) = \left(2\mathcal{L}_1^1 - 1\right)\mathcal{L}_1^1$$

Corner nodes

$N_1 = \left(2\mathcal{L}_1 - 1\right)\mathcal{L}_1$

$N_2 = \left(2\mathcal{L}_2 - 1\right)\mathcal{L}_2$

$N_3 = \left(2\mathcal{L}_3 - 1\right)\mathcal{L}_3$

Mid-side nodes

$N_4 = 4\mathcal{L}_1\mathcal{L}_2$

$N_5 = 4\mathcal{L}_2\mathcal{L}_3$

$N_6 = 4\mathcal{L}_3\mathcal{L}_1$

Cubic
(Fig.1.8(c))

$$\mathcal{L}_0^3(x) = \frac{1}{2}\left(3\mathcal{L}_1^1 - 1\right)\left(3\mathcal{L}_1^1 - 2\right)\mathcal{L}_0^1$$

$$\mathcal{L}_1^3(x) = \frac{9}{2}\mathcal{L}_1^1\left(3\mathcal{L}_0^1 - 1\right)\mathcal{L}_0^1$$

$$\mathcal{L}_2^3(x) = \frac{9}{2}\mathcal{L}_1^1\left(3\mathcal{L}_1^1 - 1\right)\mathcal{L}_0^1$$

$$\mathcal{L}_3^3(x) = \frac{1}{2}\mathcal{L}_1^1\left(3\mathcal{L}_1^1 - 1\right)\left(3\mathcal{L}_1^1 - 2\right)$$

Corner nodes

$N_1 = \frac{1}{2}\left(3\mathcal{L}_1 - 1\right)\left(3\mathcal{L}_1 - 2\right)\mathcal{L}_1$

$N_2 = \frac{1}{2}\left(3\mathcal{L}_2 - 1\right)\left(3\mathcal{L}_2 - 2\right)\mathcal{L}_2$

$N_3 = \frac{1}{2}\left(3\mathcal{L}_3 - 1\right)\left(3\mathcal{L}_3 - 2\right)\mathcal{L}_3$

Edge nodes

$N_4 = \frac{9}{2}\mathcal{L}_1\mathcal{L}_2\left(3\mathcal{L}_1 - 1\right)$

$N_5 = \frac{9}{2}\mathcal{L}_1\mathcal{L}_2\left(3\mathcal{L}_2 - 1\right)$

$N_6 = \frac{9}{2}\mathcal{L}_2\mathcal{L}_3\left(3\mathcal{L}_2 - 1\right)$

$N_7 = \frac{9}{2}\mathcal{L}_2\mathcal{L}_3\left(3\mathcal{L}_3 - 1\right)$

$N_8 = \frac{9}{2}\mathcal{L}_3\mathcal{L}_1\left(3\mathcal{L}_3 - 1\right)$

$N_9 = \frac{9}{2}\mathcal{L}_3\mathcal{L}_1\left(3\mathcal{L}_1 - 1\right)$

Internal node

$N_{10} = 27\mathcal{L}_1\mathcal{L}_2\mathcal{L}_3$ †

† No direct comparison possible with Lagrange polynomial.

which when written out in 'long hand' yields

$$
\left.
\begin{aligned}
\mathcal{L}_i + \mathcal{L}_j + \mathcal{L}_m &= 1 \\
x &= \mathcal{L}_i x_i + \mathcal{L}_j x_j + \mathcal{L}_m x_m \\
y &= \mathcal{L}_i y_i + \mathcal{L}_j y_j + \mathcal{L}_m y_m
\end{aligned}
\right\}
\qquad (1.27\text{(b)})
$$

The first equation in Eqn.(1.27(b)) shows that only two of the variables are in fact independent, while the second and the third equations give the relationship between the Cartesian coordinates and the area coordinates.

With the help of the area coordinates, it is now possible to establish a whole family of triangular plane stress elements, some of which are shown in Fig.1.8. Unlike the Lagrange elements, this family of triangular elements are used much more often in practice, especially for the linear and the quadratic elements.

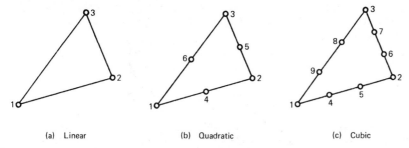

(a) Linear   (b) Quadratic   (c) Cubic

Fig.1.8 Family of triangular elements

Based on our knowledge of Lagrange polynomials, it is now a simple matter to construct the shape functions for the elements given in Fig. 1.8 and the process is shown in Table 1.3.

1.8 Choice of displacement functions

In general, a good displacement function should satisfy the following criteria.

(1) The displacement function, if given in the form of a simple polynomial, must have the same number of polynomial constants as the total number of degrees of freedom of the element. Care should be exercised in deciding on the degrees of freedom at a node, since on the one hand they must not be allowed to go below a minimum number, which ensures that the relevant deformation pattern will be adequately represented, and on the other hand they should not be so large as to make the computational procedure complex or cumbersome, even though increasing the number of degrees of freedom will usually mean improving the accuracy of an element. For example, while it is only necessary to have nodal displacements as degrees of freedom for two- or three-dimensional elasticity problems, it is, however, also necessary to use rotations (first derivative of the displacements) as additional degrees of freedom for beam and plate elements which resist external loads through bending. Degrees of freedom representing nodal strains can also be used if so desired.

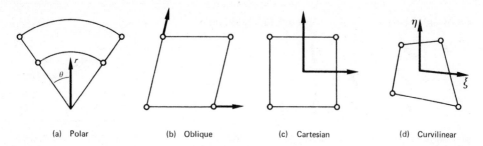

(a) Polar   (b) Oblique   (c) Cartesian   (d) Curvilinear

Fig.1.9 Plane elements of different shapes

A point worth noting is that different elements require the use of different coordinate systems. For example, the displacement functions for the sector element (Fig.1.9(a)) and parallelogrammic element (Fig. 1.9(b)) should be given in polar coordinates and oblique coordinates respectively, and not in Cartesian coordinates as for the rectangular element in Fig.1.9(c)). A much more difficult case is the quadrilateral element shown in Fig.1.9(d), where one is tempted to apply the same function $A_1 + A_2x + A_3y + A_4xy$. Unfortunately, while this function is a good one for the rectangular element, it is certainly not suitable for the quadrilateral and the correct displacement function should be $A_1 + A_2\xi + A_3\eta + A_4\xi\eta$ with ξ and η being a set of curvilinear coordinate axes.

It is not always an easy task to select the correct polynomial terms for a given number of degrees of freedom, and one of the best examples can be found in the case of a triangular bending element (Fig.1.10(a)) with nine degrees of freedom although there are altogether ten terms for a complete cubic polynomial (Fig.1.10(b)). The choice here is not at all obvious and in fact various researchers have demonstrated that there is no satisfactory solution available. The first successful nine-degrees-of-freedom triangular element was formulated through the use of area coordinates and not directly as a simple polynomial function.

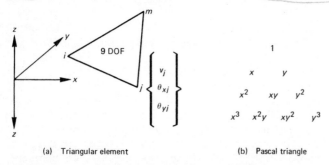

(a) Triangular element   (b) Pascal triangle

Fig.1.10 A triangular bending element

(2) In most cases, the displacement function should be balanced with respect to all coordinate axes, since most elements are classified as general purpose elements applicable to all types of problems. In this context, it will not be advisable to delete either the x^2y or xy^2 term for the triangular element shown in Fig.1.10(a) in order to satisfy criterion (1) discussed above, since such an element will yield solutions with accuracy biased in one direction.

It is important to note, however, that the statement in this criterion in no way excludes the use of special purpose elements with unbalanced displacement functions, and a good example of such unbalanced elements is the rectangular beam type plane stress element listed in Table 1.1. As can be seen, the variation of u (axial displacement of beam) and v (vertical displacement of beam) are linear and cubic respectively along the axial direction, and this in fact agrees well with the actual behaviour of a beam. Other examples having wide applications are the finite strips developed by Cheung [6] and others, in which a polynomial in one direction is used in conjunction with transcendental functions in another direction.

(3) The displacement function must allow the element to undergo rigid body movements without at the same time causing any internal strain.

For displacement functions given in terms of simple polynomials, it is only necessary to make certain that the first few terms at the top of the Pascal triangle have been included. For example, the terms, ℓ, x, y in Fig.1.9(b) would represent adequately the rigid body movements of the triangular bending element shown in Fig.1.10(a), since they would allow the element to take up another position in space on a flat inclined plane.

For displacements given in terms of shape functions it is fairly difficult to identify the rigid body terms directly, and the following two procedures are used.

(i) For elements with only scalar degrees of freedom, such as the constant strain triangle and the isoparametric elements which will be presented in detail in later chapters, the element will undergo translation or rotation during rigid body movements. For translation in any one direction all the nodes should have the same displacement f_r. This means that the displacement function given in Eqn.(1.16(b)) becomes

$$f_r = \left(\sum N_i \right) f_r \quad \text{or}$$
$$\sum N_i = 1 \tag{1.28(a)}$$

For a rigid body rotation θ around the origin of the coordinate axes means that the u displacement at any point will be equal to θy and, therefore, we have $u_i = \theta y$. Eqn.(1.16(b)) can be modified to

$$f = u = \theta y = \sum N_i \theta y_i$$

or

$$y = \sum N_i y_i \tag{1.28(b)}$$

Similarly, it can be shown that, by examining the v displacements,

$$x = \sum N_i x_i \tag{1.28(c)}$$

Thus, criterion (3) will be satisfied if all the nodal shape functions of an element add up to unity and if the x and y coordinates can be expressed as the sum of the products of the shape functions and the relevant nodal coordinates.

As an example, the shape functions for the quadratic plane stress triangular element of Fig.1.8(b) are

$$[N] = \left[\left(2\mathscr{L}_1 - 1\right)\mathscr{L}_1, \; \left(2\mathscr{L}_2 - 1\right)\mathscr{L}_2, \; \left(2\mathscr{L}_3 - 1\right)\mathscr{L}_3, \right.$$
$$\left. 4\mathscr{L}_1\mathscr{L}_2, \; 4\mathscr{L}_2\mathscr{L}_3, \; 4\mathscr{L}_3\mathscr{L}_1 \right] \tag{1.28(d)}$$

and

$$\sum N_i = \left(2\mathscr{L}_1 - 1\right)\mathscr{L}_1 + \left(2\mathscr{L}_2 - 1\right)\mathscr{L}_2 + \left(2\mathscr{L}_3 - 1\right)\mathscr{L}_3$$
$$+ 4\mathscr{L}_1\mathscr{L}_2 + 4\mathscr{L}_2\mathscr{L}_3 + 4\mathscr{L}_3\mathscr{L}_1$$

Taking into account that $\mathscr{L}_1 + \mathscr{L}_2 + \mathscr{L}_3 = 1$ (*see* Eqn.(1.27(b))), and using the relationship to eliminate $\mathscr{L}_1$

$$\sum N_i = \left(1 - 2\mathscr{L}_2 - 2\mathscr{L}_3\right)\left(1 - \mathscr{L}_2 - \mathscr{L}_3\right) + \left(2\mathscr{L}_2 - 1\right)\mathscr{L}_2$$
$$+ \left(2\mathscr{L}_3 - 1\right)\mathscr{L}_3 + 4\left(1 - \mathscr{L}_2 - \mathscr{L}_3\right)\mathscr{L}_2 + 4\mathscr{L}_2\mathscr{L}_3$$
$$+ 4\mathscr{L}_3\left(1 - \mathscr{L}_2 - \mathscr{L}_3\right)$$

$$= 1$$

To prove that Eqn.(1.28(b)) and Eqn.(1.28(c)) are also valid we take into account that $x_4 = (x_1 + x_2)/2$, $x_5 = (x_2 + x_3)/2$ and $x_6 = (x_3 + x_1)/2$. Therefore,

$$\sum N_i x_i = \left(2\mathscr{L}_1 - 1\right)\mathscr{L}_1 x_1 + \left(2\mathscr{L}_2 - 1\right)\mathscr{L}_2 x_2 + \left(2\mathscr{L}_3 - 1\right)\mathscr{L}_3 x_3$$
$$+ 4\mathscr{L}_1\mathscr{L}_2(x_1 + x_2)/2 + 4\mathscr{L}_2\mathscr{L}_3(x_2 + x_3)/2$$
$$+ 4\mathscr{L}_2\mathscr{L}_3(x_3 + x_1)/2$$
$$= x_1\mathscr{L}_1\left(2\mathscr{L}_1 + 2\mathscr{L}_2 + 2\mathscr{L}_3 - 1\right) + x_2\mathscr{L}_2\left(2\mathscr{L}_1 + 2\mathscr{L}_2 + 2\mathscr{L}_3 - 1\right)$$
$$+ x_3\mathscr{L}_3\left(2\mathscr{L}_1 + 2\mathscr{L}_2 + 2\mathscr{L}_3 - 1\right)$$
$$= x_1\mathscr{L}_1 + x_2\mathscr{L}_2 + x_3\mathscr{L}_3$$

Thus, by virtue of Eqn.(1.27(b)),

$$x = \sum N_i x_i$$

Similarly,

$$y = \sum N_i y_i$$

(ii) For elements with additional degrees of freedom such as rotations, an indirect test procedure is adopted.
(a) Give an element some sort of rigid body movements and identity the corresponding nodal displacements.
(b) Substitute these nodal displacements into the displacement function and find out whether the function will in fact be reduced to the assumed rigid body terms.

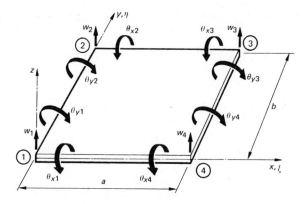

Fig.1.11 Rectangular plate element

For example, the displacement function of a non-conforming rectangular bending element [7] (Fig.1.11) is

$$w = [N]\{\delta\} \tag{1.29(a)}$$

in which the shape functions are

$$[N] = \Big[\big(1 - \xi\eta - (3 - 2\xi)\,\xi^2(1 - \eta)\,(1 - \xi)\,(3 - 2\eta)\,\eta^2\big), \quad (1 - \xi)\,\eta(1 - \eta)^2 b,$$

$$- \,\xi(1 - \xi)^2(1 - \eta)\,a, \quad \big((1 - \xi)\,(3 - 2\eta)\,\eta^2 + \xi(1 - \xi)\,(1 - 2\xi)\,\eta\big),$$

$$- \,(1 - \xi)\,(1 - \eta)\,\eta^2 b, \quad - \,\xi(1 - \xi)^2\eta a, \quad (3 - 2\xi)\,\xi^2\eta - \xi\eta(1 - \eta)\,(1 - 2\eta)\ ,$$

$$- \,\xi(1 - \eta)\,\eta^2 b, \quad (1 - \xi)\,\xi^2\eta a, \quad \big((3 - 2\xi)\,\xi^2(1 - \eta) + \xi\eta(1 - \eta)\,(1 - 2\eta)\big),$$

$$\xi\eta(1 - \eta)^2 b, \quad (1 - \xi)\,\xi^2(1 - \eta)\,a \Big] \tag{1.29(b)}$$

and the displacement parameters are

$$\{\delta\} = \big[w_1,\ \theta_{x1},\ \theta_{y1},\ w_2\ \ldots\ \theta_{y4}\big]^T \tag{1.29(c)}$$

in which $\xi = x/a$ and $\eta = y/b$

For a rigid body movement we would expect the element to take up the position of an inclined plane in space, i.e.

$$w = A_1 + A_2 x + A_3 y$$

in which case the nodal displacements would be

$$w_1 = A_1$$

$$w_2 = A_1 + A_3 b$$

$$w_3 = A_1 + A_2 a + A_3 b$$

$$w_4 = A_1 + A_2 a$$

$$\theta_{x1} = \theta_{x2} = \theta_{x3} = \theta_{x4} = \frac{\partial w}{\partial y} = A_3$$

$$\theta_{y1} = \theta_{y2} = \theta_{y3} = \theta_{y4} = \frac{\partial w}{\partial x} = - A_2$$

As a result of this Eqn.(1.29) is reduced to

$$w = A_1 \left\{ \left(1 - \xi\eta - (3 - 2\xi)\xi^2(1 - \eta) - (1 - \xi)(3 - 2\eta)\eta^2\right) + \left((1 - \xi)(3 - 2\eta)\eta^2 \right. \right.$$
$$+ \xi(1 - \xi)(1 - 2\xi)\eta\big) + \big((3 - 2\xi)\xi^2\eta - \xi\eta(1 - \eta)(1 - 2\eta)\big)$$
$$\left. + \big((3 - 2\xi)\xi^2(1 - \eta) + \xi\eta(1 - \eta)(1 - 2\eta)\big) \right\}$$
$$+ A_2 \left\{ \big((3 - 2\xi)\xi^2\eta - \xi\eta(1 - \eta)(1 - 2\eta)\big) a \right.$$
$$+ \big((3 - 2\xi)\xi^2(1 - \eta) + \xi\eta(1 - \eta)(1 - 2\eta)\big) a - \big(- \xi(1 - \xi)^2(1 - \eta)a\big)$$
$$\left. - \big(- \xi(1 - \xi)^2\eta a\big) - \big((1 - \xi)\xi^2\eta a\big) - \big((1 - \xi)\xi^2(1 - \eta)a\big) \right\}$$
$$+ A_3 \left\{ \big((1 - \xi)(3 - 2\eta)\eta^2 + \xi(1 - \xi)(1 - 2\xi)\eta\big) b \right.$$
$$+ \big((3 - 2\xi)\xi^2(1 - \eta) + \xi\eta(1 - \eta)(1 - 2\eta)\big) b + \big((1 - \xi)\eta(1 - \eta)^2 b\big)$$
$$\left. + \big(- (1 - \xi)(1 - \eta)\eta^2 b\big) + \big(- \xi(1 - \eta)\eta^2 b\big) + \xi\eta(1 - \eta)^2 b \right\}$$
$$= A_1 + A_2\xi a + A_3\eta b$$
$$= A_1 + A_2 x + A_3 y$$

Therefore it is concluded that rigid body displacement terms have been included in the shape functions.

(4) The displacement function must be able to represent a state of constant strain since this is the expected outcome if the elements are made smaller and smaller so that any smooth strain curve will be represented by a series of steps in the limit.

For displacement functions given in terms of simple polynomials, the solution is again quite simple, and it is only necessary to retain all the terms which are connected with strains or curvatures. For example, it is wrong to eliminate the xy term in the displacement function given in Fig.1.10(b) which represents the constant term for the twist $(\partial^2 w/\partial x \partial y)$ expression. Without such a constant term, the twist will always be zero at the origin where $x = 0$ and $y = 0$.

For displacements given in terms of shape functions, the same procedure outlined in criterion (3(ii)) is used with the modification that the set of nodal displacements should correspond to some constant strain modes such as pure bending or direct compression.

As an example, consider the biaxial compression mode on the rectangular plane stress element considered previously. The corresponding nodal displacements are $u_i = - u_j = - u_k = u_m = c$ and $- v_i = v_j = v_k = v_m = d$ and, substituting these values into Eqn.(1.28), we have

$$u = \frac{1}{4}\{(1 - \xi)(1 - \eta) - (1 + \xi)(1 - \eta) - (1 + \xi)(1 - \eta) + (1 - \xi)(1 + \eta)\}c$$
$$= - cx$$
$$v = \frac{1}{4}\{- (1 - \xi)(1 - \eta) - (1 + \xi)(1 - \eta) + (1 + \xi)(1 - \eta) + (1 - \xi)(1 + \eta)\}d$$
$$= - dy$$

Since $\varepsilon_x = \partial u/\partial x = - c$ and $\varepsilon_y = \partial v/\partial y = - d$, we can satisfy ourselves that the constant strain mode is indeed present in the shape functions.

A more complicated example involves the quadratic plane stress triangular element discussed previously, in which the displacement functions are

$$u = (2\mathcal{L}_1 - 1)\mathcal{L}_1 u_1 + (2\mathcal{L}_2 - 1)\mathcal{L}_2 u_2 + (2\mathcal{L}_3 - 1)\mathcal{L}_3 u_3$$
$$+ 4\mathcal{L}_1\mathcal{L}_2 u_4 + 4\mathcal{L}_2\mathcal{L}_3 u_5 + 4\mathcal{L}_3\mathcal{L}_1 u_6$$

$$v = (2\mathcal{L}_1 - 1)\mathcal{L}_1 v_1 + (2\mathcal{L}_2 - 1)\mathcal{L}_2 v_2 + (2\mathcal{L}_3 - 1)\mathcal{L}_3 v_3$$
$$+ 4\mathcal{L}_1\mathcal{L}_2 v_4 + 4\mathcal{L}_2\mathcal{L}_3 v_5 + 4\mathcal{L}_3\mathcal{L}_1 v_6 \qquad (1.30(a))$$

For a general constant strain state, the displacement field must be linear and, therefore,

$$u = B_1 + B_2 x + B_3 y$$
$$v = B_4 + B_5 x + B_6 y \qquad (1.30(b))$$

and the corresponding nodal displacements are obtained by substituting into Eqn.(1.30) the appropriate nodal coordinates x_1, x_2, x_3, $x_4 = (x_1 + x_2)/2$, $x_5 = (x_2 + x_3)/2$, $x_6 = (x_3 + x_1)/2$, etc. Eqn.(1.30(a)) can now be written as (*vide* Eqn.(1.27(b)))

$$u = (2\mathcal{L}_1 - 1)\mathcal{L}_1(B_1 + B_2 x_1 + B_3 y_1) + (2\mathcal{L}_2 - 1)\mathcal{L}_2(B_1 + B_2 x_2 + B_3 y_2)$$
$$+ (2\mathcal{L}_3 - 1)\mathcal{L}_3(B_1 + B_2 x_3 + B_3 y_3) + 4\mathcal{L}_1\mathcal{L}_2\left(B_1 + B_3\frac{x_1 + x_2}{2} + B_3\frac{y_1 + y_2}{2}\right)$$
$$+ 4\mathcal{L}_2\mathcal{L}_3\left(B_1 + B_2\frac{x_2 + x_3}{2} + B_3\frac{y_2 + y_3}{2}\right) + 4\mathcal{L}_3\mathcal{L}_1\left(B_1 + B_2\frac{x_3 + x_1}{2} + B_3\frac{y_3 + y_1}{2}\right)$$

$$= \left\{2(\mathcal{L}_1 + \mathcal{L}_2 + \mathcal{L}_3)^2 - (\mathcal{L}_1 + \mathcal{L}_2 + \mathcal{L}_3)\right\}B_1$$
$$+ \left\{\left(2x_1\mathcal{L}_1(\mathcal{L}_1 + \mathcal{L}_2 + \mathcal{L}_3) - x_1\mathcal{L}_1\right) + \left(2x_2\mathcal{L}_2(\mathcal{L}_1 + \mathcal{L}_2 + \mathcal{L}_3) - x_2\mathcal{L}_2\right)\right.$$
$$+ \left(2x_3\mathcal{L}_3(\mathcal{L}_1 + \mathcal{L}_2 + \mathcal{L}_3) - x_3\mathcal{L}_3\right)\right\}B_2 + \left\{\left(2y_1\mathcal{L}_1(\mathcal{L}_1 + \mathcal{L}_2 + \mathcal{L}_3) - y_1\mathcal{L}_1\right)\right.$$
$$+ \left(2y_2\mathcal{L}_2(\mathcal{L}_1 + \mathcal{L}_2 + \mathcal{L}_3) - y_2\mathcal{L}_2\right) + \left(2y_3\mathcal{L}_3(\mathcal{L}_1 + \mathcal{L}_2 + \mathcal{L}_3) - y_3\mathcal{L}_3\right)\right\}B_3$$

$$= B_1 + \left(x_1\mathcal{L}_1 + x_2\mathcal{L}_2 + x_3\mathcal{L}_3\right)B_2 + \left(y_1\mathcal{L}_1 + y_2\mathcal{L}_2 + y_3\mathcal{L}_3\right)B_3$$
$$= B_1 + xB_2 + yB_3$$

and

$$v = B_4 + xB_5 + yB_6$$

Once again the constant strain states are found to be present in the shape functions.

(5) The displacement functions should satisfy the compatibility conditions along common boundaries between adjacent elements. To be more
specific, there should be at least compatibility of displacements for
two- and three-dimensional elasticity problems, and for plate bending
problems these would be an additional requirement in the compatibility
of the first partial derivatives of the displacements.

The satisfaction of criteria (4) and (5) will guarantee the convergence of the strain energy of the system to the correct level. However,
in practice criterion (5) has often been partially relaxed to produce
non-conforming elements which could be quite well-behaved, but with no
monotonic convergence characteristics.

1.9 Formulation of stiffness matrix and load matrix

Once the displacement function has been determined, it is a fairly
straightforward matter to obtain all the strains and stresses within
the element and to formulate the stiffness matrix and consistent load
matrices and in this section the principle of minimum total potential
energy will be used in the formulation. It should be noted that a consistent load matrix contains equivalent nodal forces which represent
the action of external distributed loads or body forces, and the magnitude of these nodal forces will vary with the displacement functions
used for the element. Naturally the resultant of all the forces will
remain unaltered.

1.9.1 DISPLACEMENT FUNCTIONS

A displacement function given in terms of simple polynomials can be
written in matrix form as

$$\{f\} = [R]\{A\} \tag{1.31}$$

in which $\{f\}$ may have one or more components (e.g. $[u,v]^T$ for plane
stress element and $\{w\}$ for bending element), $[R]$ is a function of the
coordinates and $\{A\}$ is a vector of polynomial constants.

It is necessary to express the displacement function in terms of nodal
displacement parameters $\{\delta\}$ and the relationships between the parameters
and the polynomial constants are obtained by simply substituting into
Eqn. (1.31) the nodal coordinates of the elements. Thus

$$\{\delta\} = [C]\{A\} \tag{1.32}$$

from which

$$\{A\} = [C]^{-1}\{\delta\} \tag{1.33}$$

and finally

$$\{f\} = [R][C]^{-1}\{\delta\}$$

$$= [N]\{\delta\} = \sum [N_i]\{\delta_i\} \tag{1.34}$$

which is equivalent to saying that the displacements are now given in
shape function form after carrying out the transformation process.

1.9.2 STRAINS

The strains are obtained through making appropriate differentiations of the displacement function with respect to the relevant coordinate variables x or y. Thus,

$$\{\varepsilon\} = [B]\{\delta\} \tag{1.35}$$

in which $[B]$ is called the strain matrix.

As an example, in the case of a plane stress element,

$$\{\varepsilon\} = \left\{ \begin{array}{c} \varepsilon_x \\ \\ \varepsilon_y \\ \\ \gamma_{xy} \end{array} \right\} = \left\{ \begin{array}{c} \dfrac{\partial u}{\partial x} \\ \\ \dfrac{\partial v}{\partial y} \\ \\ \dfrac{\partial u}{\partial y} + \dfrac{\partial v}{\partial x} \end{array} \right\} \tag{1.36}$$

and, in the case of a bending element,

$$\{\varepsilon\} = \left\{ \begin{array}{c} \chi_x \\ \\ \chi_y \\ \\ 2\chi_{xy} \end{array} \right\} = \left\{ \begin{array}{c} -\dfrac{\partial^2 w}{\partial x^2} \\ \\ -\dfrac{\partial^2 w}{\partial y^2} \\ \\ 2\dfrac{\partial^2 w}{\partial x \partial y} \end{array} \right\} \tag{1.37}$$

1.9.3 STRESSES

The stresses are related to the strains by

$$\{\sigma\} = [D]\{\varepsilon\}$$

$$= [D][B]\{\delta\} \tag{1.38}$$

The matrix $[D]$ is often referred to as the elasticity or property matrix. For a plane stress element with isotropic properties

$$[D] = \frac{E}{1-\nu^2} \begin{bmatrix} 1 & \nu & 0 \\ \nu & 1 & 0 \\ 0 & 0 & \dfrac{1-\nu}{2} \end{bmatrix} \tag{1.39(a)}$$

while for orthotropic properties

$$[D] = \begin{bmatrix} \dfrac{E_x}{1-\nu_x\nu_y} & \dfrac{\nu_x E_y}{1-\nu_x\nu_y} & 0 \\[3mm] \dfrac{\nu_x E_y}{1-\nu_x\nu_y} & \dfrac{E_y}{1-\nu_x\nu_y} & 0 \\[3mm] 0 & 0 & G \end{bmatrix}$$ (1.39(b))

1.9.4 MINIMIZATION OF TOTAL POTENTIAL ENERGY

(a) Strain energy
The strain energy of a linear elastic body is given by

$$U = \frac{1}{2}\int \{\varepsilon\}^T\{\sigma\}\mathrm{d(vol)}$$ (1.40(a))

which after substituting for $\{\varepsilon\}$ and $\{\sigma\}$ from Eqn.(1.35) and Eqn.(1.38) becomes

$$U = \frac{1}{2}\int \{\delta\}^T[B]^T[D][B]\{\delta\}\mathrm{d(vol)}$$ (1.40(b))

(b) Potential energy
The potential energy due to distributed loads $\{q\}$ can be written as

$$W = -\int \{f\}^T\{q\}\mathrm{d(vol)}$$
$$\quad = -\int \{\delta\}^T[N]^T\{q\}\mathrm{d(vol)}$$ (1.41)

Eqn.(1.41) can be degenerated into an area integral or line integral in case of surface loads or line loads. For a concentrated load, integration is no longer necessary and the expression is reduced to that of a load multiplied by the corresponding displacement.

(c) Total potential energy
From Eqn.(1.4) the total potential energy is given by

$$\phi = U + W$$
$$\quad = \frac{1}{2}\int \{\delta\}^T[B]^T[D][B]\{\delta\}\mathrm{d(vol)} - \int \{\delta\}^T[N]^T\{q\}\mathrm{d(vol)}$$ (1.42)

(d) Minimization procedure
Referring to Eqn.(1.10), the minimization procedure requires that

$$\left\{\frac{\partial\phi}{\partial\{\delta\}}\right\} = \{0\}$$ (1.43)

Therefore, if Eqn.(1.42) is substituted into Eqn.(1.43) and the partial differentiation is carried out, then

$$\left\{\frac{\partial \phi}{\partial \{\delta\}}\right\} = \int [B]^T[D][B]\{\delta\}\mathrm{d(vol)} - \int [N]^T\{q\}\mathrm{d(vol)} = \{0\} \tag{1.44}$$

or

$$[k]\{\delta\} - \{P\} = 0 \tag{1.45}$$

in which the stiffness matrix

$$[k] = \int [B]^T[D][B]\mathrm{d(vol)} \tag{1.46}$$

and the consistent load matrix

$$\{P\} = \int [N]^T\{q\}\mathrm{d(vol)} \tag{1.47}$$

A great deal of integration work has to be performed in Eqns.(1.46) and (1.47) and, while it is possible to carry out the integration in closed form for simple cases, it is advisable to use numerical integration procedure such as Gaussian quadrature for the majority of cases.

1.10 Bar element stiffness matrix rederived

Consider the bar element of Fig.1.1 which has two nodes and one degree of freedom at each node. With the total number of degrees of freedom equal to two, in accordance with criterion (1) in Section 1.8 there should only be two polynomial constants. The stiffness matrix, which has been given in Eqn.(1.13), will be rederived in accordance with the finite element formulation procedure.

(i) *Displacement function in polynomial form*

$$\{f\} = \{u\} = A_1 + A_2 x$$

Therefore,

$$[R] = [1 \quad x]$$

$$\{A\} = \begin{Bmatrix} A_1 \\ A_2 \end{Bmatrix} \tag{1.48}$$

(ii) *Nodal displacement parameter-polynomial constant relationship*
Substituting the nodal coordinates $x = 0$ and $x = \ell$ into Eqn.(1.48) one after the other, we have

$$u_1 = A_1 + A_2(0)$$

$$u_2 = A_1 + A_2(\ell)$$

Therefore,

$$[C] = \begin{bmatrix} 1 & 0 \\ 1 & \ell \end{bmatrix} \qquad [C]^{-1} = \frac{1}{\ell}\begin{bmatrix} \ell & 0 \\ -1 & 1 \end{bmatrix}$$

(iii) *Displacement function in shape function form*
Referring to Eqn.(1.34), the shape functions

$$[N] = [R][C]^{-1}$$

$$= \begin{bmatrix} 1 & x \end{bmatrix} \frac{1}{\ell} \begin{bmatrix} \ell & 0 \\ -1 & 1 \end{bmatrix}$$

$$= \begin{bmatrix} \left(1 - \frac{x}{\ell}\right) & \frac{x}{\ell} \end{bmatrix} = \begin{bmatrix} N_1 & N_2 \end{bmatrix} \tag{1.49}$$

The above expression in fact corresponds to the linear Lagrange polynomial given in Table 1.2. Thus it is quite easy to formulate the
displacements in shape function directly in this case.

(iv) *Strain-nodal displacement parameter relationship*
Only axial strain is present in this case and

$$\{\varepsilon\} = \varepsilon_x = \frac{du}{dx} = \begin{bmatrix} \dfrac{dN_1}{dx} & \dfrac{dN_2}{dx} \end{bmatrix} \begin{Bmatrix} u_1 \\ u_2 \end{Bmatrix}$$

$$= \begin{bmatrix} -\dfrac{1}{\ell} & \dfrac{1}{\ell} \end{bmatrix} \begin{Bmatrix} u_1 \\ u_2 \end{Bmatrix}$$

Therefore,

$$[B] = \begin{bmatrix} -\dfrac{1}{\ell} & \dfrac{1}{\ell} \end{bmatrix} \tag{1.50}$$

(v) *Stress-strain relationship*
The relationship is simply given by Hooke's Law, and

$$\{\sigma\} = \sigma_x = E\varepsilon_x$$

Therefore

$$[D] = [E] \tag{1.51}$$

(vi) *Stiffness matrix*
From Eqn.(1.46), the stiffness matrix is

$$[k] = \int [B]^T [D][B] d(\text{vol})$$

$$= \int_A d(\text{area}) \int_0^\ell [B]^T [D][B] dx$$

$$= A \int_0^\ell \begin{Bmatrix} \dfrac{-1}{\ell} \\ \dfrac{1}{\ell} \end{Bmatrix} [E] \begin{bmatrix} \dfrac{-1}{\ell} & \dfrac{1}{\ell} \end{bmatrix} dx$$

$$= \frac{EA}{\ell} \begin{bmatrix} 1 & -1 \\ -1 & 1 \end{bmatrix} \tag{1.52}$$

A quick check will show that the two matrices given in Eqns.(1.13) and (1.52) are identical, and this demonstrates the validity of the finite element formulation.

1.11 Beam element stiffness matrix rederived

Consider now the beam element shown in Fig.1.2. The two nodal deflections will obviously only provide a rigid body translation for the element, and in order for bending action to occur, additional rotational degrees of freedom must be incorporated. Thus there are altogether four degrees of freedom which would define a cubic polynomial uniquely. The stiffness matrix of the beam element, which was given in Eqn.(1.15), will also be rederived in accordance with the finite element procedure.

(i) *Displacement function in polynomial form*

$$\{f\} = \{v\} = A_1 + A_2 x + A_3 x^2 + A_4 x^3$$

i.e.

$$[R] = \begin{bmatrix} 1 & x & x^2 & x^3 \end{bmatrix}$$

$$\{A\} = \begin{bmatrix} A_1 & A_2 & A_3 & A_4 \end{bmatrix}^T \tag{1.53}$$

(ii) *Nodal displacement parameter-polynomial constant relationship*
Before proceeding any further it is necessary first of all to obtain an expression for the rotation θ from the deflection v.

$$\theta = \frac{dv}{dx} = A_2 + 2A_3 x + 3A_4 x^2 \tag{1.54}$$

If the nodal coordinates are now substituted into Eqn.s(1.53) and (1.54), it is possible to establish the following equations (*vide* Eqn.(1.32)).

$$v_1 = A_1 + A_2(0) + A_3(0)^2 + A_4(0)^3$$

$$\theta_1 = \quad A_2 \quad + 2A_3(0) \quad + 3A_4(0)^2$$

$$v_2 = A_1 + A_2(\ell) + A_3(\ell)^2 + A_4(\ell)^3$$

$$\theta_2 = \quad A_2 \quad + 2A_3(\ell) \quad + 3A_4(\ell)^2$$

Therefore,

$$[C] = \begin{bmatrix} 1 & 0 & 0 & 0 \\ 0 & 1 & 0 & 0 \\ 1 & \ell & \ell^2 & \ell^3 \\ 0 & 1 & 2\ell & 3\ell^2 \end{bmatrix} \qquad [C]^{-1} = \begin{bmatrix} 1 & 0 & 0 & 0 \\ 0 & 1 & 0 & 0 \\ -3/\ell^2 & -2/\ell & 3/\ell^2 & -1/\ell \\ 2/\ell^3 & 1/\ell^2 & -2/\ell^3 & 1/\ell^2 \end{bmatrix}$$

(iii) *Displacement function in shape function form*

$$[N] = [R][C]^{-1} = \begin{bmatrix} 1 & x & x^2 & x^3 \end{bmatrix} \begin{bmatrix} 1 & 0 & 0 & 0 \\ 0 & 1 & 0 & 0 \\ -3/\ell^2 & -2/\ell & 3/\ell^2 & -1/\ell \\ 2/\ell^3 & 1/\ell^2 & -2/\ell^3 & 1/\ell^2 \end{bmatrix}$$

$$= \begin{bmatrix} \left(1 - \dfrac{3x^2}{\ell^2} + \dfrac{2x^3}{\ell^3}\right), & x\left(1 - \dfrac{2x}{\ell} + \dfrac{x^2}{\ell^2}\right), & \left(\dfrac{3x^2}{\ell^2} - \dfrac{2x^3}{\ell^3}\right), & x\left(\dfrac{x^2}{\ell^2} - \dfrac{x}{\ell}\right) \end{bmatrix}$$

(1.55)

The expression in Eqn.(1.55) is in fact the same as the first-order Hermitian polynomial given in Eqn.(1.22)

(iv) *Strain-nodal displacement parameter relationship*
For a beam element, the strain corresponds to the curvature and, therefore,

$$\{\varepsilon\} = -\frac{d^2v}{dx^2} = \begin{bmatrix} \left(\dfrac{6}{\ell^2} - \dfrac{12x}{\ell^3}\right), & \left(\dfrac{4}{\ell} - \dfrac{6x}{\ell^2}\right), & \left(\dfrac{-6}{\ell^2} + \dfrac{12x}{\ell^3}\right), & \left(\dfrac{2}{\ell} - \dfrac{6x}{\ell^2}\right) \end{bmatrix}\{\delta\}$$

or

$$[B] = \begin{bmatrix} \left(\dfrac{6}{\ell^2} - \dfrac{12x}{\ell^3}\right), & \left(\dfrac{4}{\ell} - \dfrac{6x}{\ell^2}\right), & \left(\dfrac{-6}{\ell^2} + \dfrac{12x}{\ell^3}\right), & \left(\dfrac{2}{\ell} - \dfrac{6x}{\ell^2}\right) \end{bmatrix} \qquad (1.56)$$

(v) *Stress-strain relationship*
The stress-strain relationship corresponds to moment-curvature relationship in the case of a beam and we have

$$\{\sigma\} \equiv M = EI\left(-\frac{d^2v}{dx^2}\right)$$

Therefore,

$$[D] \equiv EI$$

(vi) *Stiffness matrix*
With $[B]$ and $[D]$ known, it is a simple matter to work out the stiffness matrix in accordance with Eqn.(1.46), bearing in mind that the volume integral can now be reduced to a line integral.

$$[k] = \int_0^\ell [B]^T [D][B]\,dx$$

$$= EI \int_0^\ell \begin{bmatrix} \dfrac{36}{\ell^4} - \dfrac{144x}{\ell^5} + \dfrac{144x^2}{\ell^6} & & & \\[2ex] \dfrac{24}{\ell^3} - \dfrac{84x}{\ell^4} + \dfrac{72x^2}{\ell^5} & \dfrac{16}{\ell^2} - \dfrac{48x}{\ell^3} + \dfrac{36x^2}{\ell^4} & \text{Symmetrical} & \\[2ex] -\dfrac{36}{\ell^4} + \dfrac{144x}{\ell^5} - \dfrac{144x^2}{\ell^6} & \dfrac{-24}{\ell^3} + \dfrac{84x}{\ell^4} - \dfrac{72x^2}{\ell^5} & \dfrac{36}{\ell^4} - \dfrac{144x}{\ell^5} + \dfrac{144x^2}{\ell^6} & \\[2ex] \dfrac{12}{\ell^3} - \dfrac{60x}{\ell^4} + \dfrac{72x^2}{\ell^5} & \dfrac{8}{\ell^2} - \dfrac{36x}{\ell^3} + \dfrac{36x^2}{\ell^4} & \dfrac{-12}{\ell^3} + \dfrac{60x}{\ell^4} - \dfrac{72x^2}{\ell^5} & \dfrac{4}{\ell^2} - \dfrac{24x}{\ell^3} + \dfrac{36x^2}{\ell^4} \end{bmatrix} dx$$

$$= EI \begin{bmatrix} \dfrac{12}{\ell^3} & \text{Symmetrical} & & \\[2ex] \dfrac{6}{\ell^2} & \dfrac{4}{\ell} & & \\[2ex] \dfrac{-12}{\ell^3} & \dfrac{-6}{\ell^2} & \dfrac{12}{\ell^3} & \\[2ex] \dfrac{6}{\ell^2} & \dfrac{2}{\ell} & \dfrac{-6}{\ell^2} & \dfrac{4}{\ell} \end{bmatrix} \qquad (1.58)$$

As expected, the stiffness matrices given in Eqns.(1.15) and (1.58) are identical, although the formulation procedure is different.

(vi) *Consistent load matrix for uniformly distributed loading*
For a beam element, Eqn.(1.47) should be reduced to

$$\{P\} = \int_0^\ell [N]^T \{q\} dx$$

$$= \int_0^\ell \begin{Bmatrix} \left(1 - \dfrac{3x^2}{\ell^2} + \dfrac{2x^3}{\ell^3}\right) \\[2ex] x\left(1 - \dfrac{2x}{\ell} + \dfrac{x^2}{\ell^2}\right) \\[2ex] \left(3\dfrac{x^2}{\ell^2} - \dfrac{2x^3}{\ell^3}\right) \\[2ex] x\left(-\dfrac{x}{\ell} + \dfrac{x^2}{\ell^2}\right) \end{Bmatrix} q\,dx \qquad (1.59(a))$$

In the case of uniformly distributed load, q is constant and, therefore,

$$\{P\} = q \left\{ \begin{array}{c} x - \dfrac{x^3}{\ell^2} + \dfrac{x^4}{2\ell^3} \\[3mm] \dfrac{x^2}{2} - \dfrac{2}{3}\dfrac{x^3}{\ell} + \dfrac{1}{4}\dfrac{x^4}{\ell^2} \\[3mm] \dfrac{x^3}{\ell^2} - \dfrac{1}{2}\dfrac{x^4}{\ell^3} \\[3mm] \dfrac{-x^3}{3\ell} + \dfrac{1}{4}\dfrac{x^4}{\ell^2} \end{array} \right\}_{0}^{\ell} = \left\{ \begin{array}{c} \dfrac{q\ell}{2} \\[3mm] \dfrac{q\ell^2}{12} \\[3mm] \dfrac{q\ell}{2} \\[3mm] \dfrac{-q\ell^2}{12} \end{array} \right\} \qquad (1.59(b))$$

Eqn.(1.59(b)) coincides with the nodal forces obtained by reversing the fixed end forces acting on a clamped beam under uniformly distributed load.

References

1. O.C. Zienkiewicz. *Finite Element Methods in Engineering Science*. New York, McGraw-Hill Book Co., 1971.

2. B. Fraeijs de Veubeke. Upper and lower bounds in matrix structural analysis. In *Matrix Methods of Structural Analysis*. AGARD, Vol.72, 1964, pp.165-201.

3. T.H.H. Pian. Derivation of element stiffness matrices. *Journal of the American Institute of Aeronautics and Astronautics*, 2, 576-7, 1964.

4. R.W. Clough. The finite element method in structural mechanics. In *Stress Analysis* (O.C. Zienkiewicz and G.S. Holister, eds). New York, John Wiley, 1965.

5. F. Schied. *Theory and Problems of Numerical Analysis*. New York, McGraw-Hill Book Co., 1968.

6. Y.K. Cheung. *Finite Strip Method in Structural Analysis*. Oxford, Pergamon Press, 1976.

7. J.S. Przemieniecki. *Theory of Matrix Structural Analysis*. New York, McGraw-Hill Book Co., 1968.

2 Triangular Element for Plane Elasticity and a Simple Computer Program

2.1 Introduction

With the basis of the finite element formulation established in Chapter 1 and its application to one-dimensional elements fully demonstrated, it is appropriate at this stage to attend to continuum problems, and in partic- ular to two-dimensional elasticity problems [1] which were the first successful examples of the application of the finite element method.

In this chapter, the simple triangular element listed in Table 1.1 will be derived in detail, and a simple but workable computer program will be described in a step by step fashion. Because of its simplicity the triangular element is the logical element to start off with in the learn- ing process, although in no way should it be construed as being the best element available, since the isoparametric eight-node element to be discussed in the next chapter would be a very much better element from the point of computational effort and accuracy. The simple computer program is comprehensive and automatic in as much as all the finite element analysis steps of data input, stiffness assembly, node fixing, solution, stress computation and output have been included; the program is not, however, sophisticated because of its low level schemes of assembly and solution. This is done on purpose in order to keep the coding easy to understand, and the readers should refer to the next chapter for an advanced computer program.

It is possible to include both plane stress ($\sigma_z = 0$) and plane strain ($\varepsilon_z = 0$) problems into the analysis and this is discussed in the deriv- ation of the stiffness matrix. However, to avoid any unnecessary con- fusion the simple program has been written only for plane stress analysis. For plane strain analysis modified elastic constants will have to be used as data input.

2.2 Constant strain triangular element

The simple triangular element referred to in Table 1.1 and shown in Fig. 2.1 is very often called a constant strain triangle, since the strains can be proved to be constant at every point within the element. The stiffness and consistent load matrices of this element will now be derived by applying the standard equations given in Section 1.9.

It is obvious that at each node there must be at least two degrees

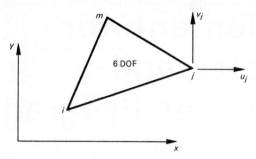

Fig.2.1 *Constant strain triangular element*

of freedom, a u displacement in the x direction and a v displacement in the y direction. Hence there would be a total of three degrees of freedom for u and v respectively, and consequently three polynomial constants for each displacement function.

2.2.1 DISPLACEMENT FUNCTION IN POLYNOMIAL FORM FROM THE PASCAL TRIANGLE

The choice of the polynomial function presents no difficulty and has already been shown in Table 1.1.

$$u = A_1 + A_2 x + A_3 y$$

$$= \begin{bmatrix} 1 & x & y \end{bmatrix} \begin{bmatrix} A_1 & A_2 & A_3 \end{bmatrix}^T$$

$$= [R]\{A\} \tag{2.1a}$$

$$v = B_1 + B_2 x + B_3 y$$

$$= [R]\{B\} \tag{2.1b}$$

Therefore

$$\{f\} = \begin{Bmatrix} u \\ v \end{Bmatrix} = \begin{bmatrix} [R] & [o] \\ [o] & [R] \end{bmatrix} \begin{Bmatrix} \{A\} \\ \{B\} \end{Bmatrix} = \begin{bmatrix} 1 & x & y & 0 & 0 & 0 \\ 0 & 0 & 0 & 1 & x & y \end{bmatrix} \begin{Bmatrix} A_1 \\ A_2 \\ A_3 \\ B_1 \\ B_2 \\ B_3 \end{Bmatrix} \tag{2.1c}$$

2.2.2 NODAL DISPLACEMENT PARAMETER–POLYNOMIAL CONSTANT RELATIONSHIP

Since both u and v have the same shape function it is only necessary to find the relationship for one of them. Substituting the x and y coordinates of the nodes into Eqn.(2.1a) one after the other, we obtain

$$u_i = A_1 + x_i A_2 + y_i A_3 \tag{2.2}$$

$$u_j = A_1 + x_j A_2 + y_j A_3 \tag{2.3}$$

$$u_m = A_1 + x_m A_2 + y_m A_3 \tag{2.4}$$

The polynomial constants can be expressed in terms of the nodal displacements by carrying out Gaussian elimination as follows.

(i)   Eliminate A_1 by performing Eqn.(2.3) - Eqn.(2.2) and Eqn.(2.4) - Eqn.(2.2),

$$u_j - u_i = (x_j - x_i)A_2 + (y_j - y_i)A_3 \tag{2.5}$$

$$u_m - u_i = (x_m - x_i)A_2 + (y_m - y_i)A_3 \tag{2.6}$$

(ii)   Eliminate A_2 by performing the calculations of Eqn.(2.6) $\times$ $(x_j - x_i)$ - Eqn.(2.5) $\times$ $(x_m - x_i)$,

$$(x_m - x_j)u_i + (x_i - x_m)u_j + (x_j - x_i)u_m$$

$$= \left\{(x_j - x_i)(y_m - y_i) - (x_m - x_i)(y_j - y_i)\right\}A_3$$

or

$$A_3 = \frac{1}{2\Delta}(x_{mj}u_i + x_{im}u_j + x_{ji}u_m) \tag{2.7}$$

where $x_{mj} = (x_m - x_j)$ and $y_{jm} = (y_j - y_m)$, etc. and Δ is in fact the area of triangle ijm and can be computed, *vide* Eqn.(1.25b).

(iii) Compute A_2 by substituting Eqn.(2.7) into Eqn.(2.5),

$$A_2 = \frac{1}{2\Delta}(y_{jm}u_i + y_{mi}u_j + y_{ij}u_m) \tag{2.8}$$

(iv)   Compute A_1 by substituting Eqn.(2.7) and Eqn.(2.8) into Eqn.(2.2),

$$A_1 = \frac{1}{2\Delta}\left[(x_j y_m - x_m y_j)u_i + (x_m y_i - x_i y_m)u_j + (x_i y_j - x_j y_i)u_m\right] \tag{2.9}$$

Writing Eqns.(2.9), (2.8) and (2.7) in matrix form,

$$\begin{Bmatrix} A_1 \\ A_2 \\ A_3 \end{Bmatrix} = \frac{1}{2\Delta} \begin{bmatrix} x_j y_m - x_m y_j & x_m y_i - x_i y_m & x_i y_j - x_j y_i \\ y_{jm} & y_{mi} & y_{ij} \\ x_{mj} & x_{im} & x_{ji} \end{bmatrix} \begin{Bmatrix} u_i \\ u_j \\ u_m \end{Bmatrix} \tag{2.10}$$

The expressions in Eqn.(2.10) can be simplified considerably by introducing the notations given in Eqn.(1.25) and Eqn.(2.10) now becomes

$$\begin{Bmatrix} A_1 \\ A_2 \\ A_3 \end{Bmatrix} = \frac{1}{2\Delta} \begin{bmatrix} a_i & a_j & a_m \\ b_i & b_j & b_m \\ c_i & c_j & c_m \end{bmatrix} \begin{Bmatrix} u_i \\ u_j \\ u_m \end{Bmatrix} \tag{2.11}$$

and a similar equation can be written for the v displacements:

$$\begin{Bmatrix} B_1 \\ B_2 \\ B_3 \end{Bmatrix} = \frac{1}{2\Delta} \begin{bmatrix} a_i & a_j & a_m \\ b_i & b_j & b_m \\ c_i & c_j & c_m \end{bmatrix} \begin{Bmatrix} v_i \\ v_j \\ v_m \end{Bmatrix} \tag{2.12}$$

In finite element analysis, the displacement parameters at each node are always grouped together (for example u_i, v_i, w_i, θ_{xi}, θ_{yi}, θ_{zi}) for ease of assembly, and therefore the displacement parameter vector would be

$$\{\delta\} = \left[u_i,\ v_i,\ u_j,\ v_j,\ u_m,\ v_m\right] \tag{2.13}$$

and the $[C]^{-1}$ matrix, unlike in previous cases, must now be constructed from Eqns. (2.11) and (2.12).

$$[C]^{-1} = \frac{1}{2\Delta}
\begin{bmatrix}
a_i & 0 & a_j & 0 & a_m & 0 \\
b_i & 0 & b_j & 0 & b_m & 0 \\
c_i & 0 & c_j & 0 & c_m & 0 \\
0 & a_i & 0 & a_j & 0 & a_m \\
0 & b_i & 0 & b_j & 0 & b_m \\
0 & c_i & 0 & c_j & 0 & c_m
\end{bmatrix} \tag{2.14}$$

2.2.3 DISPLACEMENT FUNCTIONS IN SHAPE FUNCTION FORM

From Eqn. (1.34) and Eqn. (2.1c),

$$\{f\} = \begin{Bmatrix} u \\ v \end{Bmatrix} = \begin{bmatrix} [R] & [0] \\ [0] & [R] \end{bmatrix} [C]^{-1}\{\delta\} = [N]\{\delta\} \tag{2.15}$$

Therefore the shape functions

$$[N] = \frac{1}{2\Delta}\left[
\begin{array}{c|c|c}
a_i + b_i x + c_i y & 0 & a_j + b_j x + c_j y \\
0 & a_i + b_i x + c_i y & 0 \\
\end{array}\right.$$

$$\left.\begin{array}{c|c|c}
0 & a_m + b_m x + c_m y & 0 \\
a_i + b_j x + c_j y & 0 & a_m + b_m x + c_m y
\end{array}\right]$$

$$= \begin{bmatrix}
\mathscr{L}_i & 0 & \mathscr{L}_j & 0 & \mathscr{L}_m & 0 \\
0 & \mathscr{L}_i & 0 & \mathscr{L}_j & 0 & \mathscr{L}_m
\end{bmatrix}$$

in which $\mathscr{L}_i$, $\mathscr{L}_j$, $\mathscr{L}_m$ are the area coordinates and thus the shape functions are really nothing but the linear functions shown in Table 1.3, which can be written down directly in the first place.

2.2.4 STRAIN-NODAL DISPLACEMENT PARAMETER RELATIONSHIP

The strain components for a two-dimensional elasticity problem are two direct strains ε_x, ε_y and shear strain γ_{xy} and they are given by

$$\varepsilon_x = \frac{\partial u}{\partial x}, \qquad \varepsilon_y = \frac{\partial v}{\partial y}, \qquad \gamma_{xy} = \frac{\partial u}{\partial y} + \frac{\partial v}{\partial x} \tag{2.16}$$

Substituting the displacement functions given in Eqn.(2.15) and carrying out the appropriate partial differentiations required in Eqn.(2.16),

$$\varepsilon_x = \frac{1}{2\Delta}(b_i u_i + b_j u_j + b_m u_m)$$

$$\varepsilon_y = \frac{1}{2\Delta}(c_i v_i + c_j v_j + c_m v_m)$$

$$\gamma_{xy} = \frac{1}{2\Delta}(c_i u_i + b_i v_i + c_j u_j + b_j v_j + c_m u_m + b_m v_m)$$

and the strain matrix $[B]$ is established as

$$[B] = \frac{1}{2\Delta}\begin{bmatrix} b_i & 0 & b_j & 0 & b_m & 0 \\ 0 & c_i & 0 & c_j & 0 & c_m \\ c_i & b_i & c_j & b_j & c_m & b_m \end{bmatrix} \tag{2.17}$$

2.2.5 STRESS-STRAIN RELATIONSHIP

The two cases of plane stress and plane strain will be discussed separately since the stress-strain relationships are different for the two cases. Only isotropic materials will be considered here.

(a) Plane stress
In plane stress problems only the three stress components (σ_x, σ_y, τ_{xy}) within the x-y plane are present, the other three components (σ_z, τ_{yz}, τ_{zx}) being equal to zero. The stress-strain equations for three-dimensional elasticity are thus reduced to

$$\sigma_x = \frac{E}{(1 + \nu)(1 - 2\nu)}\left[(1 - \nu)\varepsilon_x + \nu\varepsilon_y + \nu\varepsilon_z\right]$$

$$\sigma_y = \frac{E}{(1 + \nu)(1 - 2\nu)}\left[\nu\varepsilon_x + (1 - \nu)\varepsilon_y + \nu\varepsilon_z\right]$$

$$\sigma_z = 0 = \frac{E}{(1 + \nu)(1 - 2\nu)}\left[\nu\varepsilon_x + \nu\varepsilon_y + (1 - \nu)\varepsilon_z\right]$$

$$\tau_{xy} = \frac{E}{2(1 + \nu)}\gamma_{xy}$$

which can be further reduced to

$$
\left.
\begin{aligned}
\sigma_x &= \frac{E}{1 - \nu^2}(\varepsilon_x + \nu\varepsilon_y) \\[2em]
\sigma_y &= \frac{E}{1 - \nu^2}(\nu\varepsilon_x + \varepsilon_y) \\[2em]
\tau_{xy} &= \frac{E}{2(1 + \nu)}\,\gamma_{xy}
\end{aligned}
\right\}
\qquad (2.18)
$$

In matrix form, Eqn. (2.18) becomes

$$
\left\{
\begin{array}{c}
\sigma_x \\[1.5em]
\sigma_y \\[1.5em]
\tau_{xy}
\end{array}
\right\}
=
\left[
\begin{array}{ccc}
\dfrac{E}{1 - \nu^2} & \dfrac{\nu E}{1 - \nu^2} & 0 \\[1.5em]
\dfrac{\nu E}{1 - \nu^2} & \dfrac{E}{1 - \nu^2} & 0 \\[1.5em]
0 & 0 & \dfrac{E}{2(1 + \nu)}
\end{array}
\right]
\left\{
\begin{array}{c}
\varepsilon_x \\[1.5em]
\varepsilon_y \\[1.5em]
\gamma_{xy}
\end{array}
\right\}
\qquad (2.19)
$$

and thus the elasticity matrix $[D]$ has been established.

(b) Plane strain
Unlike plane stress it is ε_z and not σ_z which is equal to zero in plane strain. Because of this, it is more convenient to express the strains in terms of the stresses, i.e.

$$
\varepsilon_x = \frac{1}{E}(\sigma_x - \nu\sigma_y - \nu\sigma_z)
$$

$$
\varepsilon_y = \frac{1}{E}(- \nu\sigma_x + \sigma_y - \nu\sigma_z)
$$

$$
\varepsilon_z = 0 = \frac{1}{E}(- \nu\sigma_x - \nu\sigma_y + \sigma_z)
$$

$$
\gamma_{xy} = \frac{2(1 + \nu)}{E}\,\tau_{xy}
$$

from which it can be established that

$$
\sigma_z = \nu(\sigma_x + \sigma_y)
$$

and that

$$
\left\{
\begin{array}{c}
\varepsilon_x \\[1.5em]
\varepsilon_y \\[1.5em]
\gamma_{xy}
\end{array}
\right\}
=
\frac{1}{E}
\left[
\begin{array}{ccc}
(1 - \nu^2) & -\nu(1 + \nu) & 0 \\[1.5em]
-\nu(1 + \nu) & (1 - \nu^2) & 0 \\[1.5em]
0 & 0 & 2(1 + \nu)
\end{array}
\right]
\left\{
\begin{array}{c}
\sigma_x \\[1.5em]
\sigma_y \\[1.5em]
\tau_{xy}
\end{array}
\right\}
\qquad (2.20)
$$

Finally the stress-strain relationship is arrived at by solving Eqn. (2.20) so that

$$\begin{Bmatrix} \sigma_x \\ \sigma_y \\ \tau_{xy} \end{Bmatrix} = \frac{E}{1+\nu} \begin{bmatrix} \dfrac{1-\nu}{1-2\nu} & \dfrac{\nu}{1-2\nu} & 0 \\ \dfrac{\nu}{1-2\nu} & \dfrac{1-\nu}{1-2\nu} & 0 \\ 0 & 0 & \dfrac{1}{2} \end{bmatrix} \begin{Bmatrix} \varepsilon_x \\ \varepsilon_y \\ \gamma_{xy} \end{Bmatrix} \qquad (2.21)$$

The two elasticity matrices given in Eqns. (2.19) and (2.21) have the same form and it is more convenient to represent them by the following common elasticity matrix

$$[D] = \begin{bmatrix} C_1 & C_1 C_2 & 0 \\ C_1 C_2 & C_1 & 0 \\ 0 & 0 & C_{12} \end{bmatrix} \qquad (2.22)$$

in which

$C_1 = E/(1 - \nu^2)$ and $C_2 = \nu$ for plane stress,

$C_1 = \dfrac{E(1-\nu)}{(1+\nu)(1-2\nu)}$ and $C_2 = \dfrac{\nu}{1-\nu}$ for plane strain,

and

$C_{12} = C_1(1 - C_2)/2$ for both cases.

The element stresses are given by Eqn. (1.38), which after substituting in Eqns. (2.22) and (2.17), becomes

$$\begin{Bmatrix} \sigma_x \\ \sigma_y \\ \tau_{xy} \end{Bmatrix} = \frac{1}{2\Delta} \begin{bmatrix} C_1 & C_1 C_2 & 0 \\ C_1 C_2 & C_1 & 0 \\ 0 & 0 & C_{12} \end{bmatrix} \begin{bmatrix} b_i & 0 & b_j & 0 & b_m & 0 \\ 0 & c_i & 0 & c_j & 0 & c_m \\ c_i & b_i & c_j & b_j & c_m & b_m \end{bmatrix} \begin{Bmatrix} u_i \\ v_i \\ u_j \\ v_j \\ u_m \\ v_m \end{Bmatrix}$$

$$= \frac{1}{2\Delta}
\begin{bmatrix}
C_1 b_i & C_1 C_2 c_i & C_1 b_j & C_1 C_2 c_j & C_1 b_m & C_1 C_2 c_m \\
C_1 C_2 b_i & C_1 c_i & C_1 C_2 b_j & C_1 c_j & C_1 C_2 b_m & C_1 c_m \\
C_{12} c_i & C_{12} b_i & C_{12} c_j & C_{12} b_j & C_{12} c_m & C_{12} b_m
\end{bmatrix}
\begin{Bmatrix}
u_i \\ v_i \\ u_j \\ v_j \\ u_m \\ v_m
\end{Bmatrix}$$

$$(2.23)$$

Since none of the coefficients in the stress matrix of Eqn. (2.23) are functions of x and y, it is concluded that the stresses are constant over the whole element resulting in stress discontinuities between adjacent elements. In practice, the values of the stresses are assigned to the centroid of each triangular element so that a smooth flow of stresses can be achieved throughout the whole domain under consideration.

TABLE 2.1 *Stiffness matrix of constant strain triangle*

$$[k] = \frac{t}{4\Delta}
\begin{bmatrix}
C_1 b_i^2 + C_{12} c_i^2 & & & & & \\
C_1 C_2 b_i c_i + C_{12} b_i c_i & C_1 c_i^2 + C_{12} b_i^2 & & & \text{symmetrical} & \\
C_1 b_i b_j + C_{12} c_i c_j & C_1 C_2 b_j c_i + C_{12} b_i c_j & C_1 b_j^2 + C_{12} c_j^2 & & & \\
C_1 C_2 b_i c_j + C_{12} b_j c_i & C_1 c_i c_j + C_{12} b_i b_j & C_1 C_2 b_j c_j + C_{12} b_j c_j & C_1 c_j^2 + C_{12} b_j^2 & & \\
C_1 b_i b_m + C_{12} c_i c_m & C_1 C_2 b_m c_i + C_{12} b_i c_m & C_1 b_j b_m + C_{12} c_j c_m & C_1 C_2 b_m c_j + C_{12} b_j c_m & C_1 b_m^2 + C_{12} c_m^2 & \\
C_1 C_2 b_i c_m + C_{12} b_m c_i & C_1 c_i c_m + C_{12} b_i b_m & C_1 C_2 b_j c_m + C_{12} b_m c_j & C_1 c_j c_m + C_{12} b_j b_m & C_1 C_2 b_m c_m + C_{12} b_m c_m & C_1 c_m^2 + C_{12} b_m^2
\end{bmatrix}$$

2.2.6 STIFFNESS MATRIX

The stiffness matrix is given by Eqn. (1.46) as

$$[k] = \int [B]^T [D][B] d(vol)$$

and, as indicated by Eqns. (2.17) and (2.22), both $[B]$ and $[D]$ matrices are independent of the variables x and y. Thus the stiffness matrix reduces to the following simple form

$$[k] = t\Delta [B]^T [D][B] \tag{2.24}$$

in which t is the thickness of the element and is assumed to be constant. It is a fairly simple matter to carry out the matrix multiplications given in Eqn. (2.24) and the resulting stiffness matrix is listed in Table 2.1.

In the previous discussions on a beam element it was found that the consistent load matrix corresponds to the fixed end forces (reversed in direction) of a clamped end beam under the same load. For two-dimensional finite elements under distributed loadings such fixed end forces at the nodes cannot be determined directly from classical structural mechanics and Eqn. (1.47) must be used for the computation of equivalent nodal forces. To take the simplest case of uniformly distributed forces q_x and q_y per unit area over an element, the consistent load matrix can be obtained as

$$\{P\} = \int \frac{1}{2\Delta} \begin{bmatrix} a_i + b_i x + c_i y & 0 & a_j + b_j x + c_j y & 0 \\ 0 & a_i + b_i x + c_i y & 0 & a_j + b_j x + c_j y \end{bmatrix}$$

$$\begin{bmatrix} a_m + b_m x + c_m y & 0 \\ 0 & a_m + b_m x + c_m y \end{bmatrix}^T \begin{Bmatrix} q_x \\ q_y \end{Bmatrix} dx dy \tag{2.25}$$

The three integrals involved are $\int dx dy = \Delta$, $\int x \, dx dy$ and $\int y \, dx dy$. By choosing the centroid of the triangle as the origin of the coordinate axes, the last two integrals are equal to zero, and Eqn. (2.25) is reduced to

$$\{P\} = \frac{1}{2} \begin{Bmatrix} a_i q_x \\ a_i q_y \\ a_j q_x \\ a_j q_y \\ a_m q_x \\ a_m q_y \end{Bmatrix} \tag{2.26}$$

40

A further simplification is possible by noting that if the point ℓ in Fig.1.7 coincides with the origin then $x = y = 0$, and Eqn.(1.24) will take up the form of

$$\mathcal{L}_i = A_i/\Delta = a_i/2\Delta$$
$$\mathcal{L}_j = A_j/\Delta = a_j/2\Delta$$
$$\mathcal{L}_m = A_m/\Delta = a_m/2\Delta$$

By virtue of the fact that $A_i = A_j = A_m$, it is finally established that

$$a_i = a_j = a_m = \frac{2}{3}\Delta$$

and

$$\{P\} = \frac{\Delta}{3}\begin{Bmatrix} q_x \\ q_y \\ q_x \\ q_y \\ q_x \\ q_y \end{Bmatrix} \tag{2.27}$$

2.3 Example problem for all subsequent computations

As mentioned previously, a simple computer program using the constant strain triangular element will be presented, in which the various steps of data input, stiffness generation and assembly, introduction of boundary conditions, solution, stress computation and output of results will be discussed in detail and illustrated by carrying out the numerical analysis of a simple problem. The problem chosen is that of a thick cylinder (Fig.2.2) subjected to internal pressure and zero axial force, for which an analytical solution [2] is available and the stresses are given as

Radial stress, $\qquad \sigma_r = \dfrac{a^2 p}{b^2 - a^2}\left(1 - \dfrac{b^2}{r^2}\right)$ \hfill (2.28a)

Tangential stress, $\qquad \sigma_\theta = \dfrac{a^2 p}{b^2 - a^2}\left(1 + \dfrac{b^2}{r^2}\right)$ \hfill (2.28b)

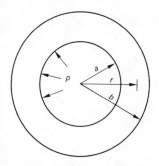

Fig.2.2 Thick cylinder under internal pressure

Before any analysis can be made it is necessary for the engineer to make two decisions, the type of elements and the mesh configuration to be used.

The first of these decisions is very often restricted by the availability of element subroutines, and engineers are forced to use lower-order elements (simple elements with few degrees of freedom and using lower-order polynomial functions) which are more readily found in program packages, in spite of the fact that some higher-order element (element of higher accuracy but with many degrees of freedom and using higher-order polynomial functions) might be preferable in view of its accuracy and ease of fitting the geometry.

The second of these decisions is linked with the first in that the order of the element used in the analysis is in general inversely proportional to the number of elements used to represent the structure, but given a certain element type and a particular problem, more elements will be required to model complex geometrical shapes and areas in which the stress gradient is steep. As there are no hard and fast rules for establishing the meshes, experience plays a very important role in the decision making. For new or important problems the standard approach is to repeat the analysis several times with successively refined meshes, and when acceptable convergence has occurred the last mesh is the standard one for all future computations for similar types of loadings.

For a number of elements, and in particular the constant strain triangular element, the accuracy of the results deteriorates rapidly with increased aspect ratio, and in general, long and narrow elements should be avoided.

To demonstrate the effect of mesh refinement on the convergence of results consider the illustrative problem of the thick cylinder given in Fig.2.2. Because of axisymmetry it is normally only required to analyse a small segment of the cylinder but for our simple program a $90°$ segment (Fig.2.3) is used here to take advantage of the symmetry conditions along the x and y axes.

A number of different mesh divisions are shown in Fig.2.3. Mesh (a) will be seen to yield bad results because of the extreme coarseness while meshes (b) and (c) will still give bad results because of insufficient refinement of mesh in the radial direction. From the results plotted in Fig.2.4 it can be seen that the results for meshes (d) and (e) are quite reasonable since the radial stress results are coincident with the analytical curve while for tangential stresses the finite element results straddle the exact curve.

42

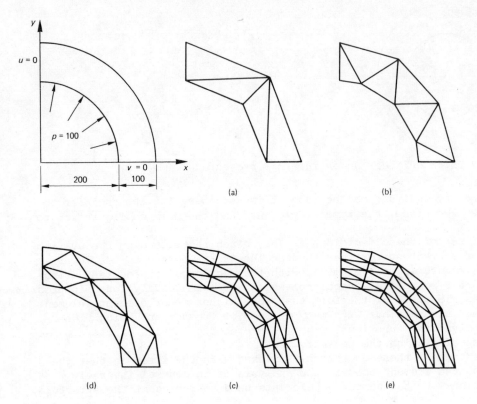

Fig.2.3 Different mesh divisions for the example problem

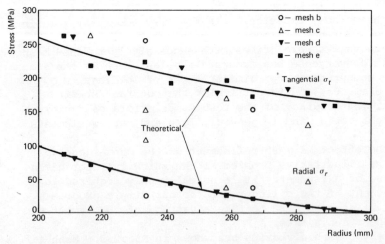

Fig.2.4 Stresses in the thick cylinder

However, in order to demonstrate explicitly all the computational steps in the execution of the finite element program mesh (a) will be used because of the comparatively small amount of computation involved.

2.4 Data preparation

After having decided on the element type and mesh division it is now possible to prepare the data for input into the program. Such data is extracted directly from the drawing of the structure in which all information concerning node and element numbering, material properties, load and boundary conditions are available (Fig.2.5). These data will be grouped under a number of headings.

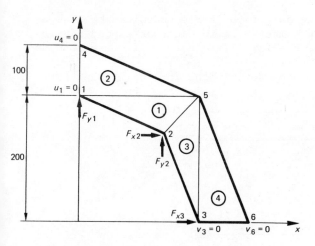

Fig.2.5 Details of example problem

2.4.1 CONTROL DATA

To make the program general and not problem dependent, it is necessary to define the main control parameters for the structure to be analysed, and this is achieved through the input of one or several control data cards at the start of the program. The control data consists of four variables and their values for the sample problem are listed below.

MAXNEL (number of elements in the structure)   = 4

MAXNOD (number of nodes in the structure)   = 6

MAXLOD (number of nodes with external applied forces) = 3

MAXFIX (number of nodes with constraints)   = 4

These variables are used to control the reading of the main body of data and for determining the limits of the main loops inside the assembly and solution part of the program. Once a program has been compiled and stored on permanent file, it can be used time after time to analyse different problems by reading in a new set of control variables without going through any recoding process.

2.4.2 ELEMENT DEFINITION

The variable MAXNEL controls the input of geometric and material properties data for all the elements. Each of the elements is defined by
three nodes and the nodes must be defined in the same order as the one
used in formulation, which is anticlockwise for the set of axes given in
Fig.2.1. Failure to comply with this will result in a non-positive
definite stiffness [3] for the element and incorrect results for the
whole structure.

In the element definition any one of the three nodes can be used as
the first node and the element number will in fact correspond to its
position in the data set, i.e. the first data card will correspond to
element number one, etc.

Referring to Fig.2.5, the element definitions for the illustrative
problem are

Element number	Node i	Node j	Node m
1	1	2	5
2	1	5	4
3	2	3	5
4	3	6	5

In accordance with the discussion given previously, it would be equally
proper to define element one as 2, 5, 1.

2.4.3 NODAL COORDINATES

The variable MAXNOD controls the input for all nodal coordinates, which
are required for the computation of stiffness, stress and load matrices
and at times for determining the orientation of an element in space.
For the illustrative problem the nodal coordinates are defined as

Node	x coordinate	y coordinate
1	0.00	200.00
2	141.42	141.42
3	200.00	0.00
4	0.00	300.00
5	212.13	212.13
6	300.00	0.00

These nodal coordinates are relative to the centre of the cylinder
with the x-y plane orientated as shown in Fig.2.5. It is allowable,
however, to use any other point as origin, and provided that there is
no rotation of the x-y plane, the various matrices will remain unchanged.

2.4.4 APPLIED LOADS

The loads which are applied to a structure must be defined in the data.
For many problems the majority of nodes do not have any loads applied to
them, and the variable MAXLOD needs only to define the number of loaded
nodes and to read in a corresponding number of data cards. By initial-
izing the load array to zero the zero values at all unloaded nodes would
be provided for automatically.

The x and y components of the loads have been calculated manually here,
although for more advanced programs they are very often computed automatically by simply specifying the pressure values and the lines or
surfaces over which they are acting.

The load data for the sample problem is listed below:

Node	x component	y component
1	2929	7071
2	10 000	10 000
3	7071	2929

Note that the loaded nodes do not have to be entered in a sequential
order since their locations in the load array are defined by the node
number preceding the load components.

2.4.5 NODAL FIXITY

The variable MAXFIX indicates the number of nodes with constraints.
Such constraints are necessary to eliminate the rigid body modes which
are associated with the element stiffness matrix as well as the stiffness
matrix of the whole structure. From statics it is well known that for
a plane problem at least three independent displacements must be
constrained.

The nodal fixity is indicated to the program by a two digit fixity
code. The first digit refers to the u displacement and the second digit
to the v displacement. A digit set to zero indicates that the associated displacement is free to move while for the case of a digit set to
unity the associated displacement is constrained.

Due to symmetry, nodes 1, 4 and 3, 6 of the sample problem are not
allowed to move in the x and y directions respectively. The corresponding data input is therefore as follows.

Node	u-displacement	v-displacement
6	0	1
1	1	0
3	0	1
4	1	0

Note that once again the order of the constrained nodes in the data set
is immaterial.

46

2.4.6 ELEMENT PROPERTIES

Before element stiffness can be calculated element material properties
must be available. For this elementary program it is assumed that all
the elements have the same isotropic properties and thickness. For more
advanced programs it is usual to allow the material properties and
element thickness to vary from one element to the next and also to allow
anisotropic material properties for each element.

   For the illustrative problem the following properties were chosen:
Young's modulus = 0.1 × 10^6 MPa, Poisson's ratio = 0.25, thickness =
1.0 mm.

2.4.7 DATA ECHO

In the majority of finite element programs a data echo is included and
the input data are printed as a matter of course, although sometimes it
is not strictly an 'echo' because of small discrepancies in the input and
output formats. The inclusion of such a data echo will allow the checking of input data easily and quickly in case of an error, and also will
serve as a means of identification if the results are stored for future
reference.

element definition

Data input coding

```
PROGRAM SIMPLE(INPUT,OUTPUT,TAPE5=INPUT,TAPE6=OUTPUT)
```

element stiffness matrix   *nodal coordinates*   *nodal loading*   *nodal restraints*

```
DIMENSION LNODS(65,3),COORDS(50,2),ALOAD(50,2),NFIX(50,2)
DIMENSION ELSTIF(6,6),SS(90,90)
```
structural stiffness matrix
```
DIMENSION SL(90),D(3,3),B(3,6),DB(3,6),DISPL(90),DISP(6)
DIMENSION STRESS(65,3)
DIMENSION REACT(2)
```
①

load vector after assembling

```
C     NVABZ IS THE NUMBER OF VARIABLES PER NODE
C     NNODZ IS THE NUMBER OF NODES PER ELEMENT



      NVABZ=2
      NNODZ=3
```
②

```
C     READ IN DATA
```
number of elements   *number of nodes*   *number of loaded nodes*   *number of constrained nodes*
```
      READ(5,1001) MAXNEL,MAXNOD,MAXLOD,MAXFIX
      WRITE(6,1002) MAXNEL,MAXNOD,MAXLOD,MAXFIX
```
③

```
C     INITIALIZE VECTORS AND MATRICES
```
number of variables (D.O.F)
```
      MAXVAR=MAXNOD*NVABZ
      DO 2 I=1,MAXNOD
      DO 1 J=1,2
```

nodal load

```
      ALOAD(I,J)=0.0
      NFIX(I,J)=0.0
    1 CONTINUE
    2 CONTINUE
      DO 4 I=1,MAXVAR
      SL(I)=0.0
      DISPL(I)=0.0
      DO 3 J=1,MAXVAR
      SS(I,J)=0.0
    3 CONTINUE
    4 CONTINUE
      DO 6 I=1,MAXNEL
      DO 5 J=1,3
      STRESS(I,J)=0.0
    5 CONTINUE
    6 CONTINUE
```

④

```
C     READ IN ELEMENT DEFINITIONS


      WRITE(6,1003)
      DO 11 NEL=1,MAXNEL
      READ(5,1004) (LNODS(NEL,I),I=1,NNODZ)
      WRITE(6,1005) NEL,(LNODS(NEL,I),I=1,NNODZ)
   11 CONTINUE
```

⑤

```
C     READ IN NODAL CO-ORDINATES


      WRITE(6,1006)
      DO 12 NOD=1,MAXNOD
      READ(5,1007) NIC,(COORDS(NIC,I),I=1,NVABZ)
      WRITE(6,1008) NIC,(COORDS(NIC,I),I=1,NVABZ)
   12 CONTINUE
```

⑥

```
C     READ IN APPLIED LOADS AND FIXITIES


      DO 14 J=1,MAXLOD
      READ(5,1010) NIC,(ALOAD(NIC,I),I=1,NVABZ)
   14 CONTINUE
      DO 15 J=1,MAXFIX
      READ(5,1011) NIC,(NFIX(NIC,I),I=1,NVABZ)
   15 CONTINUE
```

⑦

⑧

```
      WRITE(6,1009)
      DO 16 I=1,MAXNOD
      WRITE(6,1012) I,(ALOAD(I,J),J=1,NVABZ),(NFIX(I,J),J=1,NVABZ)
   16 CONTINUE


C     READ IN ELASTIC PROPERTIES
```

E   *γ*   *thickness*

```
      READ(5,1013) YM,PR,T
      WRITE(6,1014) YM,PR,T
```

⑨

(1) Dimension the program.
(2) Define the number of variables per node and the number of nodes per element.
(3) Read in the control data (number of elements, nodes, loaded nodes and fixed nodes).
(4) Initialize vectors and matrices.
(5) Read in element definitions.
(6) Read in nodal coordinates.
(7) Read in applied loads.
(8) Read in restrained nodes.
(9) Read in element properties.

Data echo of illustrative problem

```
TOTAL NUMBER OF ELEMENTS =    4
                  NODES =    6
                  LOADS =    3
               FIXITIES =    4
```

```
ELEMENT DEFINITIONS
ELEMENT   NODE1   NODE2   NODE3

   1       1       2       5
   2       1       5       4
   3       2       3       5
   4       3       6       5
```

```
NODAL CO-ORDINATES
NODE        X           Y

  1       0.000     200.000
  2     141.420     141.420
  3     200.000       0.000
  4       0.000     300.000
  5     212.130     212.130
  6     300.000       0.000
```

```
APPLIED LOADS AND FIXITIES
NODE      APPLIED LOADS          FIXITIES

  1     .293E+04   .707E+04     1     0
  2     .100E+05   .100E+05     0     0
  3     .707E+04   .293E+04     0     1
  4     0.         0.           1     0
  5     0.         0.           0     0
  6     0.         0.           0     1
```

```
MATERIAL PROPERTIES

YOUNGS MODULUS FOR MATERIAL   =   .100E+06
POISSONS RATIO FOR MATERIAL   =   .250E+00
CONSTANT ELEMENT THICKNESS    =   .100E+01

DATA COMPLETE
```

2.5 Generation of individual stiffness matrices

The structural stiffness matrix is obtained by adding together, or assembling as it is called, all the element stiffness matrices, which must therefore be computed beforehand. The simple nature of the constant strain triangle allows the element stiffness coefficients to be determined explicitly and from Table 2.1 it can be seen that for the computation of the coefficients the following data is required.

(i)   The node numbers forming the element.

(ii)   The coordinates of these nodes.

(iii) The element material properties and thickness.

In the program the three nodal numbers are retrieved from the matrix of element definitions and are called NIC1, NIC2 and NIC3. The associated nodal coordinates are X1, Y1; X2, Y2 and X3, Y3 respectively. The material properties and element thickness are input as data and the constants b_i, c_i, b_j, etc. can be calculated. It is then a simple matter to program the coefficients given in Table 2.1 and to establish the element stiffness matrix ELSTIF.

The flow chart for calculating the element stiffness is shown in Fig. 2.6 and the numerical values of the element stiffnesses for the illustrative problems are given in Section 2.6.

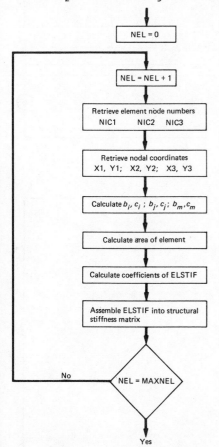

NEL = 0

NEL = NEL + 1

Retrieve element node numbers
NIC1   NIC2   NIC3

Retrieve nodal coordinates
X1, Y1; X2, Y2; X3, Y3

Calculate b_i, c_i ; b_j, c_j; b_m, c_m

Calculate area of element

Calculate coefficients of ELSTIF

Assemble ELSTIF into structural stiffness matrix

No

NEL = MAXNEL

Yes

Fig.2.6 Flow chart for stiffness generation   (NEL = element number)

50

Coding for generating element stiffness

```
C      SET UP STRUCTURAL STIFFNESS MATRIX

       EC1=YM/(1-PR*PR)
       EC2=PR
       EC12=EC1*(1-EC2)/2.0

       DO 100 NEL=1,MAXNEL

C      SET UP ELEMENT GEOMETRY

       NIC1=LNODS(NEL,1)
       NIC2=LNODS(NEL,2)
       NIC3=LNODS(NEL,3)

       X1=COORDS(NIC1,1)
       Y1=COORDS(NIC1,2)
       X2=COORDS(NIC2,1)
       Y2=COORDS(NIC2,2)
       X3=COORDS(NIC3,1)
       Y3=COORDS(NIC3,2)

       AREA=(X2*Y3-X3*Y2-X1*(Y3-Y2)+Y1*(X3-X2))/2.0

       BI=Y2-Y3
       CI=X3-X2
       BJ=Y3-Y1
       CJ=X1-X3
       BM=Y1-Y2
       CM=X2-X1

C      SET UP ELEMENT STIFFNESS MATRIX

       ELSTIF(1,1)=EC1*BI**2.0+EC12*CI**2.0
       ELSTIF(2,1)=(EC1*EC2+EC12)*BI*CI
       ELSTIF(2,2)=EC1*CI**2.0+EC12*BI**2.0
       ELSTIF(3,1)=EC1*BI*BJ+EC12*CI*CJ
       ELSTIF(3,2)=EC1*EC2*BJ*CI+EC12*BI*CJ
       ELSTIF(3,3)=EC1*BJ**2.0+EC12*CJ**2.0
       ELSTIF(4,1)=EC1*EC2*BI*CJ+EC12*BJ*CI
       ELSTIF(4,2)=EC1*CI*CJ+EC12*BI*BJ
       ELSTIF(4,3)=(EC1*EC2+EC12)*BJ*CJ
       ELSTIF(4,4)=EC1*CJ**2.0+EC12*BJ**2.0
       ELSTIF(5,1)=EC1*BI*BM+EC12*CI*CM
       ELSTIF(5,2)=EC1*EC2*BM*CI+EC12*BI*CM
       ELSTIF(5,3)=EC1*BJ*BM+EC12*CJ*CM
       ELSTIF(5,4)=EC1*EC2*BM*CJ+EC12*BJ*CM
       ELSTIF(5,5)=EC1*BM**2.0+EC12*CM**2.0
       ELSTIF(6,1)=EC1*EC2*BI*CM+EC12*BM*CI
       ELSTIF(6,2)=EC1*CI*CM+EC12*BI*BM
       ELSTIF(6,3)=EC1*EC2*BJ*CM+EC12*BM*CJ
       ELSTIF(6,4)=EC1*CJ*CM+EC12*BJ*BM
       ELSTIF(6,5)=(EC1*EC2+EC12)*BM*CM
       ELSTIF(6,6)=EC1*CM**2.0+EC12*BM**2.0
```

Handwritten annotations:

$\dfrac{E}{1-V^2}$

$\dfrac{E}{2(1+V)}$

V

plane stress ①

②

③ element nodal numbers.

④

⑤

⑥ β_i, γ_i, β_j, γ_j, β_m, γ_m

$\left(\dfrac{EV}{1-V^2}+\dfrac{E}{2(1+V)}\right)\beta_i\,\beta_i$ ⑦

```
      DO 40 I=1,5
      IP1=I+1
      DO 39 J=IP1,6
      ELSTIF(I,J)=ELSTIF(J,I)
   39 CONTINUE
   40 CONTINUE
      DO 42 I=1,6
      DO 41 J=1,6
      ELSTIF(I,J)=ELSTIF(I,J)*T/(4.0*AREA)
   41 CONTINUE
   42 CONTINUE
```

⑧

⑨

(1) Calculate coefficients of the material property matrix.
(2) Element loop.
(3) Determine the node numbers forming the element.
(4) Nodal coordinates of the element nodes.
(5) Area of the element.
(6) Calculate variables b_i, c_i, b_j, c_j, b_m, c_m.
(7) Calculate the values of the coefficients in the lower triangle of
    the stiffness matrix using Table 2.1 - with the common multiplier
    T/(4*AREA) missing.
(8) Generate the upper triangle of the element stiffness using symmetry.
(9) Multiply by Thickness/(4*AREA) to obtain the correct element
    stiffness.

2.6 Assemblage of the structural stiffness matrix

The stiffness relationship for an element has been given by Eqn.(1.45) as

$$[k]\{\delta\} = \{P\}$$

For the triangular element in question, we have

$$\begin{bmatrix} k_{ii} & k_{ij} & k_{im} \\ k_{ji} & k_{jj} & k_{jm} \\ k_{mi} & k_{mj} & k_{mm} \end{bmatrix} \begin{Bmatrix} \delta_i \\ \delta_j \\ \delta_m \end{Bmatrix} = \begin{Bmatrix} P_i \\ P_j \\ P_m \end{Bmatrix}$$

in which all the terms are in fact sub-matrices.
    Since a structure is made up of many elements, it follows that the
structural stiffness matrix is also made up of a corresponding number
of element stiffness matrices, and is given as

$$[K]\{\delta\} = \{F\}$$

where $[K]$ is the structural stiffness matrix, $\{F\}$ is a structural load
matrix with a number of load cases, $\{\delta\}$ is a matrix of all nodal
displacements.
    The process of combining the element stiffness matrices together is
called assembly, and the theory behind the assembly process will be
explained by considering the structure shown in Fig.2.7(a), in which
the triangular element definitions are:

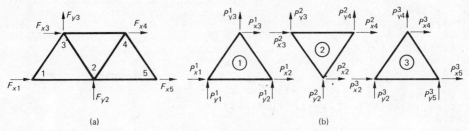

(a)   (b)

Fig.2.7 Structure and its component elements

Element number	Node i	Node j	Node m
1	1	2	3
2	2	4	3
3	2	5	4

The forces (including reactions) applied to the structure are F_{x1}, F_{y2}, F_{x3}, F_{y3}, F_{x4} and F_{x5}. The forces are in equilibrium and therefore although all of them have been shown acting in a positive sense, some of them are in fact negative. For simplication, the constraints are not shown in the figure.

   If the structure is split up into its component elements and the external forces are divided between the appropriate elements (Fig.2.7(b)) such that equilibrium and compatibility are maintained at nodal, element and structural levels, then it is obvious that from joint equilibrium

$$
\begin{aligned}
\text{Node 1} \quad & \begin{cases} p^1_{x1} = F_{x1} \\ p^1_{y1} = 0 \end{cases} \\[2mm]
\text{Node 2} \quad & \begin{cases} p^1_{x2} + p^2_{x2} + p^3_{x2} = 0 \\ p^1_{y2} + p^2_{y2} + p^3_{y2} = F_{y2} \end{cases} \\[2mm]
\text{Node 3} \quad & \begin{cases} p^1_{x3} + p^2_{x3} = F_{x3} \\ p^1_{y3} + p^2_{y3} = F_{y3} \end{cases} \\[2mm]
\text{Node 4} \quad & \begin{cases} p^2_{x4} + p^3_{x4} = F_{x4} \\ p^2_{y4} + p^3_{y4} = 0 \end{cases} \\[2mm]
\text{Node 5} \quad & \begin{cases} p^3_{x5} = F_{x5} \\ p^3_{y5} = 0 \end{cases}
\end{aligned}
\tag{2.29}
$$

From element equilibrium and displacement compatibility condition we have

$$\begin{bmatrix} k_{ii}^1 & k_{ij}^1 & k_{im}^1 \\ k_{ji}^1 & k_{jj}^1 & k_{jm}^1 \\ k_{mi}^1 & k_{mj}^1 & k_{mn}^1 \end{bmatrix} \begin{Bmatrix} \delta_1 \\ \delta_2 \\ \delta_3 \end{Bmatrix} = \begin{Bmatrix} P_1^1 \\ P_2^1 \\ P_3^1 \end{Bmatrix} \quad \text{for element 1}$$

$$\begin{bmatrix} k_{ii}^2 & k_{ij}^2 & k_{im}^2 \\ k_{ji}^2 & k_{jj}^2 & k_{jm}^2 \\ k_{mi}^2 & k_{mj}^2 & k_{mn}^2 \end{bmatrix} \begin{Bmatrix} \delta_2 \\ \delta_4 \\ \delta_3 \end{Bmatrix} = \begin{Bmatrix} P_2^2 \\ P_4^2 \\ P_3^2 \end{Bmatrix} \quad \text{for element 2} \qquad (2.30)$$

$$\begin{bmatrix} k_{ii}^3 & k_{ij}^3 & k_{im}^3 \\ k_{ji}^3 & k_{jj}^3 & k_{jm}^3 \\ k_{mi}^3 & k_{mj}^3 & k_{mn}^3 \end{bmatrix} \begin{Bmatrix} \delta_2 \\ \delta_5 \\ \delta_4 \end{Bmatrix} = \begin{Bmatrix} P_2^3 \\ P_5^3 \\ P_4^3 \end{Bmatrix} \quad \text{for element 3}$$

in which the superscripts refer to the element numbers, and k_{ii}^1, δ_1, P_1^1, etc. are submatrices involving x and y components. Substituting Eqn. (2.30) into Eqn. (2.29) gives

$$k_{ii}^1 \delta_1 + k_{ij}^1 \delta_2 + k_{im}^1 \delta_3 = P_1^1 = \begin{Bmatrix} F_{x1} \\ 0 \end{Bmatrix}$$

$$k_{ji}^1 \delta_1 + (k_{jj}^1 + k_{ii}^2 + k_{ii}^3) \delta_2 + (k_{jm}^1 + k_{im}^2) \delta_3 + (k_{ij}^2 + k_{im}^3) \delta_4$$
$$+ k_{ij}^3 \delta_5 = P_2^1 + P_2^2 + P_2^3 = \begin{Bmatrix} 0 \\ F_{y2} \end{Bmatrix}$$

$$k_{mi}^1 \delta_1 + (k_{mj}^1 + k_{mi}^2) \delta_2 + (k_{mn}^1 + k_{mn}^2) \delta_3 + k_{mj}^2 \delta_4 = P_3^1 + P_3^2 = \begin{Bmatrix} F_{x3} \\ F_{y3} \end{Bmatrix} \qquad (2.31)$$

$$(k_{ji}^2 + k_{mi}^3) \delta_2 + k_{jm}^2 \delta_3 + (k_{jj}^2 + k_{mn}^3) \delta_4 + k_{mj}^3 \delta_5 = P_4^2 + P_4^3 = \begin{Bmatrix} F_{x4} \\ 0 \end{Bmatrix}$$

$$k_{ji}^3 \delta_2 + k_{jm}^3 \delta_4 + k_{jj}^3 \delta_5 = P_5^3 = \begin{Bmatrix} F_{x5} \\ 0 \end{Bmatrix}$$

Rewriting Eqn.(2.31) in matrix form

$$\begin{bmatrix} k_{ii}^1 & k_{ij}^1 & k_{im}^1 & & \\ k_{ji}^1 & \begin{matrix} k_{jj}^1 \\ +k_{ii}^2 \\ +k_{ii}^3 \end{matrix} & \begin{matrix} k_{jm}^1 \\ +k_{im}^2 \end{matrix} & k_{ij}^2 & k_{ij}^3 \\ k_{mi}^1 & \begin{matrix} k_{mj}^1 \\ +k_{mi}^2 \end{matrix} & \begin{matrix} k_{mm}^1 \\ +k_{mm}^2 \end{matrix} & k_{mj}^2 & \\ & \begin{matrix} k_{ji}^2 \\ +k_{mi}^3 \end{matrix} & \begin{matrix} k_{jm}^2 \\ +k_{mm}^3 \end{matrix} & k_{jj}^2 & k_{mj}^3 \\ & k_{ji}^3 & k_{jm}^3 & & k_{jj}^3 \end{bmatrix} \begin{Bmatrix} \delta_1 \\ \delta_2 \\ \delta_3 \\ \delta_4 \\ \delta_5 \end{Bmatrix} = \begin{Bmatrix} F_1 \\ F_2 \\ F_3 \\ F_4 \\ F_5 \end{Bmatrix} \qquad (2.32)$$

Eqn.(2.31) is of course the expanded form of Eqn.(2.32) for the example structure.

The above process described the theoretical basis of combining the individual element equations to form the overall structural equations. For the computer an automatic assembly process yielding the same end results will be used. To effect this automatic assembling an element destination vector is established. For example the element definition vector for element 2 is (2, 4, 3) and for the simple solution subroutine to be presented in Section 2.9, this will become also the destination vector which will determine the order in which the element displacements are stored in the element displacement vector, i.e. the displacements are stored as $\begin{bmatrix} \delta_2 & \delta_4 & \delta_3 \end{bmatrix}^T$.

When the front solution technique is described in Chapter 3 the reader will observe that this destination vector is obtained through a more complicated procedure.

For the purposes of explanation only, the operation of the destination vector can be seen most easily if the vector is placed above and to the left of the element stiffness and alongside the element load vector. Thus for element 2 we have

$$\begin{matrix} & 2 & 4 & 3 & & & \\ 2 & \begin{bmatrix} k_{ii}^2 & k_{ij}^2 & k_{im}^2 \\ 4 & k_{ji}^2 & k_{jj}^2 & k_{jm}^2 \\ 3 & k_{mi}^2 & k_{mj}^2 & k_{mm}^2 \end{bmatrix} & \begin{Bmatrix} \delta_2 \\ \delta_4 \\ \delta_3 \end{Bmatrix} = \begin{Bmatrix} P_2^2 \\ P_4^2 \\ P_3^2 \end{Bmatrix} & \begin{matrix} 2 \\ 4 \\ 3 \end{matrix} \end{matrix}$$

The structural destination vector is simply a vector of the order in which the displacements are stored in the structural displacement vector. Usually it is an array in ascending order and for this example it is (1, 2, 3, 4, 5). If this structural destination vector is now placed above and to the left of the structural stiffness matrix and alongside the structural load vector the required automatic assembly process is obtained by adding any one of the 2 × 2 submatrices in the element stiffness into the structural stiffness according to the location defined by the destination vectors above and to the left of the element stiffness. For example k_{jm}^2 will be assembled into the 2 × 2 submatrix area defined by location (4,3) in the structural stiffness matrix. The row position, 4, is obtained from the vertical destination vector to the left of the element stiffness and the column position, 3, from the horizontal destination vector above the element stiffness.

In a similar manner the load vector is also assembled. For example P_4^2 has alongside it the coefficient 4 from the destination vector, and is therefore added into the structural load vector into the 2 × 1 subvector area that has alongside it a 4 from the structural destination vector.

The structural stiffness after assembly of element 2 only is shown in Eqn. (2.33).

$$
\begin{array}{ccccc}
1 & 2 & 3 & 4 & 5
\end{array}
$$

$$
\begin{array}{c}
1 \\ 2 \\ 3 \\ 4 \\ 5
\end{array}
\left[
\begin{array}{ccc}
 & & \\
k_{ii}^2 & k_{im}^2 & k_{ij}^2 \\
k_{mi}^2 & k_{mm}^2 & k_{mj}^2 \\
k_{ji}^2 & k_{jm}^2 & k_{jj}^2 \\
 & &
\end{array}
\right]
\left\{
\begin{array}{c}
\delta_1 \\ \delta_2 \\ \delta_3 \\ \delta_4 \\ \delta_5
\end{array}
\right\}
=
\left\{
\begin{array}{c}
 \\ P_2^2 \\ P_3^2 \\ P_4^2 \\
\end{array}
\right\}
\begin{array}{c}
1 \\ 2 \\ 3 \\ 4 \\ 5
\end{array}
\qquad (2.33)
$$

The reader can now check Eqn. (2.32) and confirm that the coefficients of element 2 are assembled into the structural coefficients as shown in Eqn. (2.33).

If elements 1 and 3 are assembled in the same manner Eqn. (2.32) would be reproduced exactly.

Coding for assembling structural stiffness matrix

```
C       ADD IN ELEMENT STIFFNESS INTO STRUCTURAL MATRIX


        DO 50 I=1,NNODZ
        DO 49 II=1,NVABZ
        ISTRST=(LNODS(NEL,I)-1)*NVABZ+II ──────────────────① 
        IELEMT=(I-1)*NVABZ+II ────────────────────────────②
        DO 48 J=1,NNODZ
        DO 47 JJ=1,NVABZ
        JSTRST=(LNODS(NEL,J)-1)*NVABZ+JJ ─────────────────③
        JELEMT=(J-1)*NVABZ+JJ ────────────────────────────④
        SS(ISTRST,JSTRST)=SS(ISTRST,JSTRST)+ELSTIF(IELEMT,JELEMT)
     47 CONTINUE
     48 CONTINUE
     49 CONTINUE
     50 CONTINUE
    100 CONTINUE
```

(1),(3) Destination in structural stiffness matrix (ISTRST, JSTRST) of coefficient.

(2),(4) (IELEMT, JELEMT) from the element stiffness.

Element stiffness no. 1

Destination vector

	1		2		3	
1	-.259E+05	-.118E+05	-.244E+05	.154E+05	-.148E+04	-.357E+04
	-.118E+05	.259E+05	.220E+05	-.578E+05	-.102E+05	.319E+05
2	-.244E+05	.220E+05	.642E+05	-.606E+04	-.397E+05	-.160E+05
	-.154E+05	-.578E+05	-.606E+04	.170E+06	-.929E+04	-.112E+06
3	-.148E+04	-.102E+05	-.397E+05	-.929E+04	.412E+05	.195E+05
	-.357E+04	.319E+05	-.160E+05	-.112E+06	.195E+05	.803E+05

Structural stiffness matrix after assembly of element no. 1

Structural destination vector

	1		2		3		4		5		6	
1	.259E+05	-.118E+05	-.244E+05	.154E+05	0.	0.	0.	0.	-.148E+04	-.357E+04	0.	0.
	-.118E+05	.259E+05	.220E+05	-.578E+05	0.	0.	0.	0.	-.102E+05	.319E+05	0.	0.
2	-.244E+05	.220E+05	.642E+05	-.606E+04	0.	0.	0.	0.	-.397E+05	-.160E+05	0.	0.
	-.154E+05	-.578E+05	-.606E+04	.170E+06	0.	0.	0.	0.	-.929E+04	-.112E+06	0.	0.
3	0.	0.	0.	0.	0.	0.	0.	0.	0.	0.	0.	0.
	0.	0.	0.	0.	0.	0.	0.	0.	0.	0.	0.	0.
4	0.	0.	0.	0.	0.	0.	0.	0.	0.	0.	0.	0.
	0.	0.	0.	0.	0.	0.	0.	0.	0.	0.	0.	0.
5	-.148E+04	-.102E+05	-.397E+05	-.929E+04	0.	0.	0.	0.	.412E+05	.195E+05	0.	0.
	-.357E+04	.319E+05	-.160E+05	-.112E+06	0.	0.	0.	0.	.195E+05	.803E+05	0.	0.
6	0.	0.	0.	0.	0.	0.	0.	0.	0.	0.	0.	0.
	0.	0.	0.	0.	0.	0.	0.	0.	0.	0.	0.	0.

Element stiffness no. 4

Destination vector

```
              3                        6                        5
3    .120E+06   .293E+05   -.112E+06   -.160E+05   -.828E+04   -.133E+05
3    .293E+05   .618E+05   -.929E+04   -.397E+05   -.200E+05   -.221E+05
6   -.112E+06   -.929E+04    .113E+06   -.404E+04   -.114E+04    .133E+05
6   -.160E+05   -.397E+05   -.404E+04    .428E+05    .200E+05   -.305E+04
5   -.828E+04   -.200E+05   -.114E+04    .200E+05    .943E+04    0.
5   -.133E+05   -.221E+05    .133E+05   -.305E+04    0.          .251E+05
```

Structural stiffness matrix after assembly of all four elements

Structural destination vector

```
         1                     2                     3                     4                     5                     6
1   .878E+05  .175E+05  -.244E+05  -.154E+05   0.        0.       -.397E+05  -.929E+04  -.236E+05  -.236E+05   0.        0.
1   .175E+05  .146E+06   .220E+05  -.578E+05   0.        0.       -.160E+05  -.112E+06  -.236E+05  -.236E+05   0.        0.
2  -.244E+05  .220E+05   .234E+06  -.121E+05  -.578E+05  .154E+05  0.        0.        -.152E+05  -.252E+05   0.        0.
2  -.154E+05 -.578E+05  -.121E+05   .234E+06   .220E+05 -.244E+05  0.        0.        -.252E+05  -.152E+06   0.        0.
3   0.        0.        -.578E+05   .220E+05   .146E+06  .175E+05  0.        0.        -.236E+05  -.236E+05  -.112E+06  -.160E+05
3   0.        0.         .154E+05  -.244E+05   .175E+05  .878E+05  0.        0.        -.236E+05  -.236E+05  -.929E+04  -.397E+05
4  -.397E+05 -.160E+05   0.        0.        0.        0.         .428E+05  -.404E+04  -.305E+04   .200E+05   0.        0.
4  -.929E+04 -.112E+06   0.        0.        0.        0.        -.404E+04   .113E+06   .133E+06  -.114E+04  -.133E+05  -.114E+04
5  -.236E+05 -.236E+05  -.152E+05  -.252E+05  -.236E+05 -.236E+05 -.305E+04   .133E+06   .391E+06  -.114E+04   .156E+06  -.200E+05
5  -.236E+05 -.236E+05  -.252E+05  -.152E+06  -.236E+05 -.236E+05  .200E+05  -.114E+04  -.114E+04   .428E+05   .133E+05  -.305E+04
6   0.        0.        0.        0.        -.112E+06 -.929E+04  0.        -.133E+05   .156E+06   .133E+05   .200E+05  -.114E+04
6   0.        0.        0.        0.        -.160E+05 -.397E+05  0.        -.114E+04  -.200E+05  -.305E+04  -.114E+04   .428E+05
```

2.7 Introduction of prescribed displacement conditions

The problem of nodal fixity was discussed briefly in Section 2.4. Such
fixities include prescribed displacements which very often are zero displacements at rigid supports. Non-zero displacements are prescribed
for support settlements and also for boundary points of local fine mesh
analysis in which the displacements were first obtained through a coarse
mesh analysis of a much larger domain.

There are a number of methods for prescribing displacements, and four
of them will be described below. Method 4 is the method used throughout
this text because of its simplicity in the computation of reactions,
which should be computed and summed for important problems so as to
ensure that equilibrium between external loads and reactions is satisfied.

The coding for calculating nodal reactions will be listed at the end of
Section 2.9, since reactions can be calculated only after the determination of displacements.

For the purpose of discussion it will be assumed here that a matrix of
size $N \times N$ is of the form

$$
\begin{bmatrix}
a_{11} & a_{12} & \cdots & a_{1n} & \cdots & a_{1N} \\
a_{21} & a_{22} & \cdots & a_{2n} & \cdots & a_{2N} \\
\vdots & \vdots & & \vdots & & \vdots \\
a_{n1} & a_{n2} & \cdots & a_{nn} & \cdots & a_{nN} \\
\vdots & \vdots & & \vdots & & \vdots \\
a_{N1} & a_{N2} & \cdots & a_{Nn} & \cdots & a_{NN}
\end{bmatrix}
\begin{Bmatrix}
x_1 \\ x_2 \\ \vdots \\ x_n \\ \vdots \\ x_N
\end{Bmatrix}
=
\begin{Bmatrix}
b_1 \\ b_2 \\ \vdots \\ b_n \\ \vdots \\ b_N
\end{Bmatrix}
\tag{2.34}
$$

in which x_i is a nodal displacement and b_i a nodal load, and that furthermore the displacement x_n has been prescribed to be equal to α, which can
be a zero or non-zero value.

2.7.1 METHOD 1

In this method, the row vector corresponding to the prescribed variable
is deleted while the column vector is multiplied by α and transferred
to the right-hand side. The matrix is reduced in size to $(N-1) \times (N-1)$
and takes up the form of

$$
\begin{bmatrix}
a_{11} & a_{12} & \cdots & a_{1,n-1} & a_{1,n+1} & \cdots & a_{1N} \\
a_{21} & a_{22} & \cdots & a_{2,n-1} & a_{2,n+1} & \cdots & a_{2N} \\
\vdots & \vdots & & \vdots & \vdots & & \vdots \\
a_{n-1,1} & a_{n-1,2} & \cdots & a_{n-1,n-1} & a_{n-1,n+1} & \cdots & a_{n-1,N} \\
a_{n+1,1} & a_{n+1,2} & \cdots & a_{n+1,n-1} & a_{n+1,n+1} & \cdots & a_{n+1,N} \\
\vdots & \vdots & & \vdots & \vdots & & \vdots \\
a_{N1} & a_{N2} & \cdots & a_{N,n-1} & a_{N,n+1} & \cdots & a_{NN}
\end{bmatrix}
\begin{Bmatrix}
x_1 \\ x_2 \\ \vdots \\ x_{n-1} \\ x_{n+1} \\ \vdots \\ x_N
\end{Bmatrix}
=
\begin{Bmatrix}
b_1 \\ b_2 \\ \vdots \\ b_{n-1} \\ b_{n+1} \\ \vdots \\ b_N
\end{Bmatrix}
- \alpha
\begin{Bmatrix}
a_{1n} \\ a_{2n} \\ \vdots \\ a_{n-1,n} \\ a_{n+1,n} \\ \vdots \\ a_{N,n}
\end{Bmatrix}
\tag{2.35}
$$

This method is only used in manual computation because of the advantage of having a reduced matrix. For automatic computation the efforts involved in rearranging the coefficients far outweigh the gain achieved through the reduction in matrix size except perhaps for eigenvalue problems in which core storage is critical.

For the computation of reaction corresponding to the displacement x_n, the original row vector must be stored and then multiplied with the computed displacements, i.e. the reaction is given by

$$R_n = \sum_{i=1}^{N} a_{ni} x_i - b_n \qquad (2.36)$$

For the more common case of $\alpha = 0$ it is of course not necessary to carry out the modification of the right-hand side as indicated in Eqn.(2.35).

2.7.2 METHOD 2

This method is simple to understand and is easier to implement when compared with Method 1. The diagonal coefficient corresponding to x_n is made unity and the rest of the coefficients in the row are then set equal to zero. To maintain symmetry in the matrix, the coefficients in the corresponding column are multiplied by α and transferred to the right-hand side, and after this the column (apart from the diagonal coefficient) is also set to zero. Lastly the coefficient b_n is replaced by α and the final form of the equations is shown below.

$$
\begin{bmatrix}
a_{11} & a_{12} & \cdots & a_{1,n-1} & 0 & a_{1,n+1} & \cdots & a_{1N} \\
a_{21} & a_{22} & \cdots & a_{2,n-1} & 0 & a_{2,n+1} & \cdots & a_{2N} \\
\cdot & \cdot & & \cdot & & \cdot & & \cdot \\
\cdot & \cdot & & \cdot & & \cdot & & \cdot \\
\cdot & \cdot & & \cdot & & \cdot & & \cdot \\
a_{n-1,1} & a_{n-1,2} & \cdots & a_{n-1,n-1} & 0 & a_{n-1,n+1} & \cdots & a_{n-1,N} \\
0 & 0 & & 0 & 1 & 0 & & 0 \\
a_{n+1,1} & a_{n+1,2} & \cdots & a_{n+1,n-1} & 0 & a_{n+1,n+1} & \cdots & a_{n+1,N} \\
a_{N1} & a_{N2} & \cdots & a_{N,n-1} & 0 & a_{N,n+1} & \cdots & a_{NN}
\end{bmatrix}
\begin{Bmatrix}
x_1 \\ x_2 \\ \cdot \\ \cdot \\ \cdot \\ x_{n-1} \\ x_n \\ x_{n+1} \\ x_N
\end{Bmatrix}
=
\begin{Bmatrix}
b_1 \\ b_2 \\ \cdot \\ \cdot \\ \cdot \\ b_{n-1} \\ \alpha \\ b_{n+1} \\ b_N
\end{Bmatrix}
- \alpha
\begin{Bmatrix}
a_{1n} \\ a_{2n} \\ \cdot \\ \cdot \\ \cdot \\ a_{n-1,n} \\ 0 \\ a_{n+1,n} \\ a_{N,n}
\end{Bmatrix}
$$

$$(2.37)$$

The reaction corresponding to x_n is obtained in the same way as indicated by Eqn.(2.36).

2.7.3 METHOD 3

Method 3 requires very few operations and is preferable to Method 4 for non-zero displacements and when reactions are not required. In this method the diagonal coefficient corresponding to x_n is multiplied by a very large number, say 10^{12}, and the load term b_n is replaced by $\alpha \times a_{nn} \times 10^{12}$. The modified equations are shown below.

$$
\begin{bmatrix}
a_{11} & a_{12} & \cdots & a_{1n} & \cdots & a_{1N} \\
a_{21} & a_{22} & & a_{2n} & & a_{2N} \\
\vdots & \vdots & & \vdots & & \vdots \\
a_{n1} & a_{n2} & & a_{nn} \times 10^{12} & \cdots & a_{nN} \\
\vdots & \vdots & & \vdots & & \vdots \\
a_{N1} & a_{N2} & & a_{Nn} & \cdots & a_{NN}
\end{bmatrix}
\begin{Bmatrix}
x_1 \\ x_2 \\ \vdots \\ x_n \\ \vdots \\ x_N
\end{Bmatrix}
=
\begin{Bmatrix}
b_1 \\ b_2 \\ \vdots \\ \alpha \times a_{nn} \times 10^{12} \\ \vdots \\ b_N
\end{Bmatrix}
\tag{2.38}
$$

Allowing for the fact that the coefficients in a row are approximately of the same order of magnitude, it is quite simple to see that the equation corresponding to x_n is in fact very nearly equivalent to $x_n = \alpha$. The reactions are obtained as before by storing the original equation and performing the calculations as given in Eqn. (2.36).

2.7.4 METHOD 4

This is the method that has been implemented in the simple program since the reaction can be obtained directly from the modified equation.

The method consists of adding a very large number, say 10^{50}, to the diagonal coefficients which physically corresponds to 'earthing' the structure with a very stiff spring. For a rigid support we obtain a very small displacement instead of an absolute zero and the reaction for that support can be computed directly as

Reaction = − (Big spring stiffness) × (Very small displacement)

For non-zero prescribed displacement it is only necessary to modify the right-hand side as is indicated in Method 2. The small displacement obtained in the solution must be replaced by the prescribed displacement before the displacement vector is used to calculate the element stresses.

This method will fail if the big spring stiffness is not significantly larger than the stiffness coefficients of the structure. However, with a value of 10^{50} this is unlikely to occur in practice.

The modified equations are given below for completeness.

$$
\begin{bmatrix}
a_{11} & a_{12} & \cdots & a_{1n} & \cdots & a_{1N} \\
a_{21} & a_{22} & \cdots & a_{2n} & \cdots & a_{2N} \\
\vdots & \vdots & & \vdots & & \vdots \\
a_{n1} & a_{n2} & \cdots & a_{nn}+10^{50} & \cdots & a_{nN} \\
\vdots & \vdots & & \vdots & & \vdots \\
a_{N1} & a_{N2} & \cdots & a_{Nn} & \cdots & a_{NN}
\end{bmatrix}
\begin{Bmatrix}
x_1 \\ x_2 \\ \vdots \\ x_n \\ \vdots \\ x_N
\end{Bmatrix}
=
\begin{Bmatrix}
b_1 \\ b_2 \\ \vdots \\ b_n \\ \vdots \\ b_N
\end{Bmatrix}
- \alpha
\begin{Bmatrix}
a_{1n} \\ a_{2n} \\ \vdots \\ a_{nn} \\ \vdots \\ a_{Nn}
\end{Bmatrix}
\tag{2.39}
$$

Coding for application of nodal fixities

```
C    APPLY FIXITIES

     I=0
     DO 200 NIC=1,MAXNJD
     DO 199 NVB=1,NVABZ
     I=I+1
     IF(NFIX(NIC,NVB).EQ.0) GO TO 199 ——①
     SS(I,I)=1.0E+50+SS(I,I) ————————②
199  CONTINUE
200  CONTINUE
```

(1) Check if displacement prescribed.
(2) Add big spring stiffness to impose prescribed zero displacement.

Structural stiffness matrix after fixity imposed

```
.100E+51   .175E+05  -.244E+05  -.154E+05   0.         0.        -.397E+05  -.929E+04  -.236E+05  -.236E+05   0.         0.
.175E+05   .146E+06  -.220E+05  -.578E+05   0.         0.        -.160E+05  -.112E+06  -.236E+05  -.236E+05   0.         0.
-.244E+05  -.220E+05   .234E+06  -.121E+05  -.578E+05   .154E+05   0.         0.        -.152E+06  -.252E+06  -.152E+05  -.252E+05
-.154E+05  -.578E+05  -.121E+05   .234E+06  -.220E+05  -.244E+05   0.         0.        -.244E+05  -.152E+05  -.252E+06  -.152E+06
0.         0.        -.578E+05  -.220E+05   .146E+06   .175E+05   0.         0.         .175E+05  -.236E+05  -.236E+05  -.112E+06
0.         0.         .154E+05  -.244E+05   .175E+05   .100E+51   0.         0.         .100E+51  -.236E+05  -.236E+05  -.929E+04
-.397E+05  -.160E+05   0.         0.         0.         0.         .391E+05   .156E+06  -.114E+04  -.305E+05  -.133E+05  -.200E+05
-.929E+04  -.112E+06   0.         0.         0.         0.         .156E+06   .391E+05  -.404E+04  -.114E+04  -.305E+05  -.133E+05
-.236E+05  -.236E+05  -.152E+06  -.252E+06  -.236E+05  -.236E+05  -.114E+04  -.404E+04   .113E+06  -.113E+06  -.404E+04   .160E+05
-.236E+05  -.236E+05  -.252E+05  -.152E+05  -.236E+05  -.236E+05  -.305E+05  -.114E+04  -.113E+06   .113E+06  -.929E+04  -.397E+05
0.         0.        -.112E+06  -.929E+04   0.         0.        -.133E+05  -.305E+05  -.404E+04  -.404E+04   .100E+51   0.
0.         0.        -.160E+05  -.397E+05   0.         0.        -.200E+05  -.133E+05   .160E+05  -.397E+05   0.         .293E+04
```

2.8 Load matrix

Only two types of loading will be considered for this initial elementary program:
   (i)   concentrated loads applied at nodes;
   (ii) uniformly distributed loads over an element edge.
Other types of loading such as gravity load, centrifugal loads, temperature loads, non-uniformly distributed loads, etc., will be described in Chapter 3 relating to isoparametric elements.

2.8.1 CONCENTRATED LOADS (NODAL POINT LOADS)

These are the easiest type of applied load to code in any finite element program. In the program described in this section the solution of equations is not carried out until all elements have been assembled into the structural stiffness matrix. The concentrated nodal loads can therefore be added into the structural load vector, SL, immediately after assembly of all the elements. Because there are two variables per node the load components X_N and Y_N applied to node N are added, or assembled, into the coefficients of SL associated with node N as listed below.

$$X_N \rightarrow SL(N*2-1)$$

$$Y_N \rightarrow SL(N*2)$$

X_N and Y_N are of course inputted as data.

2.8.2 UNIFORMLY DISTRIBUTED EDGE LOADS

This type of loading is somewhat more complicated to program than nodal point loads because it is usual in finite element programs to define only the magnitude of the distributed load per unit length, and the edge over which it is acting. The program then has to calculate the equivalent nodal forces, to determine the relative orientation of the edge to the global axes, and then transform the nodal forces from a direction perpendicular to the edge into components parallel to the global axes.
   To facilitate this process, it is usual to define a locally orientated set of axes $x'-y'$ which are parallel and perpendicular to the loaded edge respectively. For example if an element defined by nodes i, j, m in which edge $i-j$ is loaded then the locally defined x' axis will be directed from i to j, as shown in Fig.2.8.

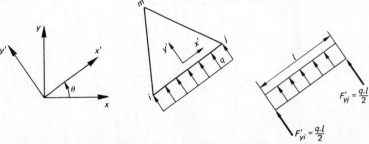

Fig.2.8 Nodal forces due to edge loads

To conform to the right-handed rule adopted in this text the y' axis must be in the direction shown. The distributed load can be either positive or negative depending on its direction.

The forces in the two systems are related to each other by

$$\begin{Bmatrix} F_x \\ F_y \end{Bmatrix} = \begin{bmatrix} \cos\theta & -\sin\theta \\ \sin\theta & \cos\theta \end{bmatrix} \begin{Bmatrix} F_x' \\ F_y' \end{Bmatrix}$$

or

$$\{F\} = [L]\{F'\}$$

in which $[L]$ is the transformation matrix between the two systems, and θ is positive as indicated in Fig.2.8.

The direction cosines are easily computed from geometrical considerations:

$$\cos\theta = (x_j - x_i)/\ell$$

$$\sin\theta = (y_j - y_i)/\ell$$

Thus the equivalent global components of forces at a node for the present example are

$$\begin{Bmatrix} F_x \\ F_y \end{Bmatrix} = \begin{bmatrix} \cos\theta & -\sin\theta \\ \sin\theta & \cos\theta \end{bmatrix} \begin{Bmatrix} 0 \\ q\ell/2 \end{Bmatrix} = (q\ell/2)\begin{Bmatrix} -\sin\theta \\ \cos\theta \end{Bmatrix}$$

Coding for assembly of applied nodal loads

```
C      ADD IN APPLIED LOADS


       DO 181 I=1,MAXNOD
       DO 180 J=1,NVABZ
       SL(2*I-2+J)=ALOAD(I,J)+SL(2*I-2+J) ─────────────①
 180 CONTINUE
 181 CONTINUE
```

(1) Add the nodal loads from the data into the load vector.

Load vector after assembly

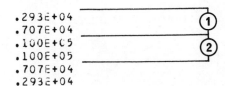

```
.293E+04 ───────────①
.707E+04
.100E+05 ───────────②
.100E+05
.707E+04
.293E+04
0.
0.
0.
0.
0.
0.
```

(1) F_x and F_y for node 1.
(2) F_x and F_y for node 2.

2.9 Solution of simultaneous equations

The most frequently used method of solving a set of simultaneous equations is Gaussian elimination [4]. This method consists of eliminating the equations one at a time so that there is a corresponding reduction in the size of the modified matrix until finally the matrix is reduced to one equation containing only one variable.

The set of eliminated equations forms a triangular matrix and is used for backsubstitution purposes. By starting at the last equation and working backwards to the first equation, one variable will be determined at each step by using the variables which are already known.

2.9.1 GAUSSIAN ELIMINATION USING A FULL MATRIX

To demonstrate the procedure for this solution technique consider a set of four simultaneous equations represented by

$$[A]\{x\} = \{b\} \tag{2.40}$$

The reduction or triangularisation of $[A]$ is carried out by computing

$$a^*_{ij} = a_{ij} - a_{is}a_{sj}/a_{ss}$$
$$b^*_i = b_i - (a_{is}/a_{ss})b_s \tag{2.41}$$

in which s refers to the equation being reduced and $i,j > s$.

The modified form of Eqn.(2.40) after the first step is

$$
\begin{bmatrix}
a_{11} & a_{12} & a_{13} & a_{14} \\
0 & a_{22}-\dfrac{a_{21}}{a_{11}}a_{12} & a_{23}-\dfrac{a_{21}}{a_{11}}a_{13} & a_{24}-\dfrac{a_{21}}{a_{11}}a_{14} \\
0 & a_{32}-\dfrac{a_{31}}{a_{11}}a_{12} & a_{33}-\dfrac{a_{31}}{a_{11}}a_{13} & a_{34}-\dfrac{a_{31}}{a_{11}}a_{14} \\
0 & a_{42}-\dfrac{a_{41}}{a_{11}}a_{12} & a_{43}-\dfrac{a_{41}}{a_{11}}a_{13} & a_{44}-\dfrac{a_{41}}{a_{11}}a_{14}
\end{bmatrix}
\begin{Bmatrix} x_1 \\ x_2 \\ x_3 \\ x_4 \end{Bmatrix}
=
\begin{Bmatrix}
b_1 \\
b_2-\dfrac{a_{21}}{a_{11}}b_1 \\
b_2-\dfrac{a_{31}}{a_{11}}b_1 \\
b_4-\dfrac{a_{41}}{a_{11}}b_1
\end{Bmatrix}
\tag{2.42}
$$

and the final form after the completion of all reduction steps is

$$
\begin{bmatrix}
a_{11} & a_{12} & a_{13} & a_{14} \\
0 & a^*_{22} & a^*_{23} & a^*_{24} \\
0 & 0 & a^{**}_{33} & a^{**}_{34} \\
0 & 0 & 0 & a^{***}_{44}
\end{bmatrix}
\begin{Bmatrix} x_1 \\ x_2 \\ x_3 \\ x_4 \end{Bmatrix}
=
\begin{Bmatrix} b_1 \\ b^*_2 \\ b^{**}_3 \\ b^{***}_4 \end{Bmatrix}
\tag{2.43}
$$

where the asterisks are used to indicate the number of times each
coefficient has been modified. The backsubstitution process is obvious
and requires no comment. The computer coding for this technique is
listed later on in this section. The whole matrix is stored in core
and the coding is straightforward and easy to understand. However, the
subroutine is inefficient in terms of core storage and computer time
because no advantage has been taken of the symmetric and banded nature
of a structural stiffness matrix.

2.9.2 GAUSSIAN ELIMINATION FOR A BANDED MATRIX OF CONSTANT WIDTH

The procedure described above uses a full matrix and is therefore
inefficient both in terms of core requirement and the number of arith-
metic operations in achieving a solution. Fortunately in many matrices
generally met in finite element problems the non-zero coefficients are
clustered near the diagonal, and outside this band of non-zero coefficients all the coefficients are of zero value. Such matrices are called
banded matrices and for most structural problems in which the nodes
have been numbered sensibly, the matrices are banded.

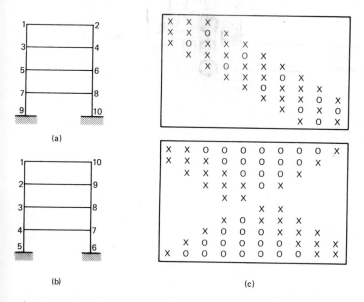

(a)

(b)

(c)

Fig.2.9 Typical frame and stiffness matrix   (X = non-zero coefficient)

In Fig.2.9(a) the nodes of a typical frame structure have been
numbered sensibly and the structural stiffness matrix can be seen to be
a fairly compact band matrix. On the other hand, by numbering the nodes
incorrectly as shown in Fig.2.9(b), the matrix is no longer banded and
the entire stiffness matrix must be used in the solution process.

The symmetry and banded nature of the matrix can be utilized in
developing a more efficient band matrix solution routine, in which only
the upper half (or lower half) of the band is stored as a rectangular
array, as is shown in Fig.2.10. It is obvious that since the core
requirement is directly proportional to the half band width which is in
turn related to the maximum difference between the nodal numbers of an

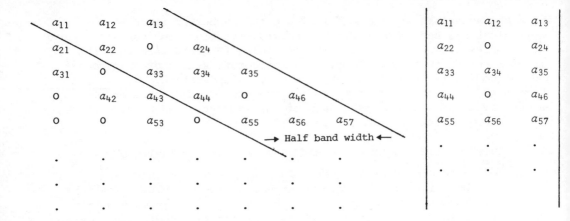

(a) Original band matrix

(b) Rectangular array
storage of upper
half of band

Fig.2.10 Storage scheme for a banded matrix

element, great care must be exercised in the numbering of nodes so that the difference between connecting nodes can be minimised.

Using the band solver routine, the storage requirement is reduced drastically since the half band width is very often only about 30 per cent of the matrix size. The execution time is also reduced because all the zero coefficients outside the band will not be operated upon.

This extra efficiency is gained at the expense of a slightly more complicated computer coding due to the shifting of the stiffness coefficients. The equivalent equation to Eqn.(2.41) is

$$a_{ij}^{*} = a_{ij} - \left(a_{s,(i-s+1)} / a_{s1}\right) a_{s,(j+1-s)} \tag{2.44}$$

with the limits

$$(s + 1) \leqslant i \leqslant (s + \text{HBW} - 1)$$

$$1 \leqslant j \leqslant \left(\text{HBW} - (i - s)\right)$$

where HBW = the half band width of the matrix. The computer listing of the band solver is given at the end of this chapter and includes modifications to the assembly, reduction and backsubstitution.

Coding for Gaussian reduction and backsubstitution

```
C       CARRY OUT GAUSSIAN REDUCTION


        M1=MAXNOD*NVABZ ───────────────────────────① 
        M2=M1-1
        DO 300 I=1,M2 ─────────────────────────────②
        II=I+1
        DO 290 K=II,M1 ────────────────────────────③
        FACT=SS(K,I)/SS(I,I) ──────────────────────④
```

```
      DO 280 J=II,M1
      SS(K,J)=SS(K,J)-FACT*SS(I,J)                    ─────┐
  280 CONTINUE                                             ⑤
      SS(K,I)=0.0                                     ─────┘
      SL(K)=SL(K)-FACT*SL(I) ─────────────────────────────⑥
  290 CONTINUE
  300 CONTINUE


  C      BACK SUBSTITUTE


      DO 400 I=1,M1 ──────────────────────────────────────⑦
      II=M1-I+1
      PIVOT=SS(II,II)                                ─────┐
      SS(II,II)=0.0                                       ⑧
      DO 350 J=II,M1                                 ─────┘
      SL(II)=SL(II)-SS(II,J)*DISPL(J)                ─────⑨
  350 CONTINUE
      DISPL(II)=SL(II)/PIVOT ─────────────────────────────⑩
  400 CONTINUE

      WRITE(6,1018)                                  ─────┐
      DO 401 I=1,MAXNOD                                    │
      WRITE(6,1019) I,DISPL(2*I-1),DISPL(2*I)             ⑪
  401 CONTINUE                                       ─────┘
```

Reduction
(1) M1 is the number of equations to be solved.
(2) Loop through each equation I in turn from 1 to M1-1 using each
    equation to reduce all equations below it.
(3) Loop to reduce all equations K below the equation I being used
    for reduction.
(4) Calculate the common multiplier a_{is}/a_{ss} from Eqn.(2.41) to avoid
    repetition of the same calculation.
(5) Modify all coefficients of equation K, the equation being reduced.
(6) Reduce the load term of equation K.

Backsubstitution
(7) Loop through all the reduced equations in turn from M1 back to 1
    using each equation to calculate one displacement.
(8) For each equation, II, store the value of the diagonal coefficient,
    PIVOT, and then set the pivot to zero in the stored equation.
(9) Subtract from the load term the sum of the products of the displace-
    ments multiplied by the equation coefficients.
(10) Divide the result of (9) by the value of the pivot to calculate the
    displacement.
(11) Print out the displacements.

Structural stiffness matrix and load vector during the reduction procedure

There is no observable change in the stiffness matrix, other than in column 1, after reduction by the first equation because of the large diagonal coefficient in this equation. After reduction by the second equation the structural stiffness matrix and load vector contain

```
.100E+51  .175E+05 -.244E+05  .154E+05  0.        0.       -.397E+05 -.929E+04 -.236E+05  0.       0.        .293E+04
0.        .146E+06  .220E+05 -.578E+05  0.        .154E+05 -.160E+06 -.112E+06 -.236E+05  0.       0.        .707E+04
0.        0.        .231E+06 -.344E+04 -.578E+05  .240E+04 -.169E+05 -.148E+06 -.288E+05  0.                 .894E+04
0.        0.        0.        .211E+06  .220E+05 -.244E+05 -.630E+04 -.443E+05 -.143E+06 -.346E+05           .128E+05
0.        0.        0.                 -.578E+05  .146E+06 -.346E+05 -.236E+05 -.112E+06 -.160E+05           .707E+05
0.        0.        0.                  .175E+05  0.                           -.236E+05 -.929E+04 -.397E+05 .293E+04
0.        0.        0.                           -.100E+51 -.163E+05 -.562E+04 -.226E+05  0.                 .771E+03
0.        0.        0.                           -.163E+05 -.274E+05 -.473E+04 -.169E+05  0.                 .542E+04
0.        0.        0.                           -.562E+04 -.473E+04  .152E+06  .428E+05 -.114E+04           .114E+04
0.        0.        0.                            .226E+05  .169E+05  .428E+05  .152E+06 -.133E+05  .200E+05 -.114E+04
0.        0.        0.                                               -.114E+04  .133E+05 -.113E+06 -.305E+04
0.        0.                                                          .200E+05 -.305E+04 -.404E+04 .100E+51
0.        0.                                                                             -.404E+04 0.
```

After complete triangularisation the result is:

```
.100E+51  .175E+05 -.244E+05  .154E+05  0.        0.       -.397E+05 -.929E+04 -.236E+05  0.       0.        .293E+04
0.        .146E+06  .220E+05 -.578E+05  0.        .154E+05 -.160E+05 -.112E+06 -.236E+05  0.       0.        .707E+04
0.        0.        .231E+06 -.344E+04 -.578E+05  .242E+05 -.169E+05 -.148E+06 -.288E+05  0.                 .894E+04
0.        0.                  .211E+06  .212E+05 -.626E+04 -.240E+04 -.143E+06 -.368E+05 -.112E+06           .129E+05
0.        0.                           .130E+06 -.238E+05 -.440E+05 -.165E+05  -.113E+05 -.368E+05           .801E+04
0.        0.                                     .100E+51 -.123E+05 -.988E+04 -.161E+05 -.350E+05 -.160E+05  .235E+04
0.        0.                                              .110E+04 -.775E+04 -.775E+04 -.188E+05 -.151E+03   .985E+03
0.        0.                                              .100E+51 -.178E+05 -.507E+04 -.106E+04  .106E+04   .692E+04
0.        0.                                                        .164E+05 -.885E+03 -.968E+04 -.746E+04   .101E+05
0.        0.                                                                  .497E+05 -.232E+04 -.928E+04 -.188E+05 .143E+05
0.        0.                                                                           .440E+05  .310E+05 -.357E+04 .465E+04
0.                                                                                               .110E+05 -.145E+05  .400E+04
0.                                                                                                        .100E+51
```

Program output for nodal displacements

```
NODAL DISPLACEMENTS


                 X COMP          Y COMP


        1        .00000          .45815
        2        .35490          .35490
        3        .45815          .00000
        4        .00000          .42169
        5        .29581          .29581
        6        .42169          .00000
```

Once the displacements have been determined the reactions at fixed nodes can be determined by multiplying the small displacement in the direction of the fixity by the big spring stiffness.

Coding for calculating nodal reactions

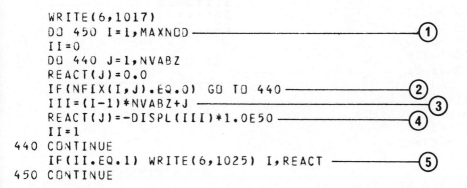

```
      WRITE(6,1017)
      DO 450 I=1,MAXNOD
      II=0
      DO 440 J=1,NVABZ
      REACT(J)=0.0
      IF(NFIX(I,J).EQ.0) GO TO 440
      III=(I-1)*NVABZ+J
      REACT(J)=-DISPL(III)*1.0E50
      II=1
440   CONTINUE
      IF(II.EQ.1) WRITE(6,1025) I,REACT
450   CONTINUE
```

(1) Loop through each of the nodes in turn.
(2) Check each fixity code for earthed displacement.
(3) Locate the associated displacement when an earthed displacement is located.
(4) Calculate the reaction.
(5) Print out reaction.

Program output for nodal reactions

```
NODAL REACTIONS


NODE      X COMP         Y COMP


    1   -.160E+05     0.
    3    0.           -.160E+05
    4   -.400E+04     0.
    6    0.           -.400E+04
```

70

2.10 Calculation of element stresses

Once the displacements have been obtained it is possible to compute the stresses of each element in turn by using Eqn.(1.38)

$$\{\sigma\} = [D][B]\{\delta\}$$

The matrices $[D]$ and $[B]$ for the constant strain element are given by Eqn.(2.22) and Eqn.(2.17) respectively and the computer coding for the whole process is listed below. The stress output for the example problem is given at the end of the computer listing.

Coding for calculating element stress

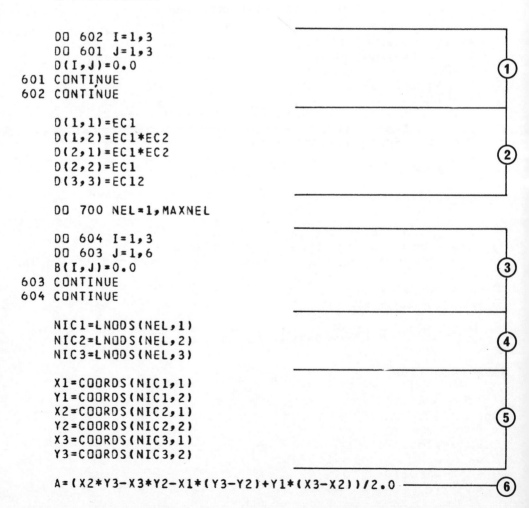

```
        CALCULATE STRESSES


        WRITE(6,1020)


        DO 602 I=1,3
        DO 601 J=1,3
        D(I,J)=0.0
  601 CONTINUE
  602 CONTINUE


        D(1,1)=EC1
        D(1,2)=EC1*EC2
        D(2,1)=EC1*EC2
        D(2,2)=EC1
        D(3,3)=EC12


        DO 700 NEL=1,MAXNEL


        DO 604 I=1,3
        DO 603 J=1,6
        B(I,J)=0.0
  603 CONTINUE
  604 CONTINUE


        NIC1=LNODS(NEL,1)
        NIC2=LNODS(NEL,2)
        NIC3=LNODS(NEL,3)


        X1=COORDS(NIC1,1)
        Y1=COORDS(NIC1,2)
        X2=COORDS(NIC2,1)
        Y2=COORDS(NIC2,2)
        X3=COORDS(NIC3,1)
        Y3=COORDS(NIC3,2)


        A=(X2*Y3-X3*Y2-X1*(Y3-Y2)+Y1*(X3-X2))/2.0
```

① ② ③ ④ ⑤ ⑥

```
      B(1,1)=(Y2-Y3)/(2.0*A)
      B(1,3)=(Y3-Y1)/(2.0*A)
      B(2,4)=(X1-X3)/(2.0*A)
      B(2,2)=(X3-X2)/(2.0*A)
      B(1,5)=(Y1-Y2)/(2.0*A)
      B(2,6)=(X2-X1)/(2.0*A)
      DO 605 I=1,3
      B(3,2*I-1)=B(2,2*I)
      B(3,2*I)=B(1,2*I-1)
  605 CONTINUE

      DO 608 I=1,3
      DO 607 J=1,6
      DB(I,J)=0.0
      DO 606 K=1,3
      DB(I,J)=DB(I,J)+D(I,K)*B(K,J)
  606 CONTINUE
  607 CONTINUE
  608 CONTINUE

      DO 609 I=1,3
      II=2*LNODS(NEL,I)-1
      JJ=2*LNODS(NEL,I)
      DISP(2*I-1)=DISPL(II)
      DISP(2*I)=DISPL(JJ)
  609 CONTINUE

      DO 620 I=1,3
      DO 619 J=1,6
      STRESS(NEL,I)=STRESS(NEL,I)+DB(I,J)*DISP(J)
  619 CONTINUE
  620 CONTINUE
      WRITE(6,1021) NEL,(STRESS(NEL,J),J=1,3)
  700 CONTINUE
```

⑦ ⑧ ⑨ ⑩ ⑪

(1) Initialize the material property matrix $[D]$.
(2) Set up the $[D]$ matrix.
(3) Initialize the strain matrix $[B]$.
(4) Determine the node numbers forming the element.
(5) Nodal coordinates of the element nodes.
(6) Area of the element.
(7) Set up the first two rows of the $[B]$ matrix.
(8) Set up the third row of $[B]$ using the values of the derivatives calculated in (7).
(9) Multiply $[D][B]$.
(10) Set up the element displacement vector DISP( ).
(11) Calculate the element stresses by multiplying the result of (9) by DISP.

Program output for element stresses

ELEMENT STRESSES

	SIGMA X-X	SIGMA Y-Y	TAU X-Y
1	161.18303	32.83678	-125.05258
2	139.02280	-1.70267	-29.77643
3	32.83678	161.18303	-125.05258
4	-1.70267	139.02280	-29.77643

Stress calculation for element no. 1

$$[D] = \begin{bmatrix} .107E+06 & .267E+05 & 0. \\ .267E+05 & .107E+06 & 0. \\ 0. & 0. & .400E+05 \end{bmatrix}$$

$$[B] = \begin{bmatrix} -.500E-02 & 0. & .858E-03 & 0. & .414E-02 & 0. \\ 0. & .500E-02 & 0. & -.150E-01 & 0. & .100E-01 \\ .500E-02 & -.500E-02 & -.150E-01 & .858E-03 & .100E-01 & .414E-02 \end{bmatrix}$$

$$[D][B] = \begin{bmatrix} -.533E+03 & .133E+03 & .915E+02 & -.400E+03 & .442E+03 & .267E+03 \\ -.133E+03 & .533E+03 & .229E+02 & -.160E+04 & .110E+03 & .107E+04 \\ .200E+03 & -.200E+03 & -.600E+03 & .343E+02 & .400E+03 & .166E+03 \end{bmatrix}$$

$$\{\delta\} = \begin{Bmatrix} .160E-45 \\ .458E+00 \\ .355E+00 \\ .355E+00 \\ .296E+00 \\ .296E+00 \end{Bmatrix}$$

$$\{\delta\} = [D][B]\{\delta\} = \begin{Bmatrix} 161.18303 \\ 32.83678 \\ -125.05258 \end{Bmatrix}$$

The theoretical stress vector is $\begin{Bmatrix} 143.14 \\ 16.86 \\ -136.28 \end{Bmatrix}$. The discrepancy is due to the coarseness of the mesh.

Format statements

```
1001 FORMAT(4I5)
1002 FORMAT(1H1,5X,
    1*TOTAL NUMBER OF ELEMENTS =*,I5,/,25X,*NODES =*,I5,
    1/,25X,*LOADS =*,I5,/,22X,*FIXITIES =*,I5)
1003 FORMAT(///,5X,*ELEMENT DEFINITIONS*,/,5X,*ELEMENT*,2X,*NODE1*,2X,
    1*NODE2*,2X,*NODE3*,/)
1004 FORMAT(3I5)
1005 FORMAT(5X,I5,3(2X,I5))
1006 FORMAT(///,5X,*NODAL CO-ORDINATES*,/,5X,*NODE*,6X,*X*,9X,*Y*,/)
1007 FORMAT(I5,2(5X,F10.6))
1008 FORMAT(5X,I4,2F10.3)
1010 FORMAT(I5,2(5X,F10.3))
1011 FORMAT(3I5)
1009 FORMAT(///,5X,*APPLIED LOADS AND FIXITIES*,/,5X,*NODE*,5X,
    1*APPLIED LOADS*,8X,*FIXITIES*,/)
1012 FORMAT(5X,I4,3X,2E10.3,2I5)
1013 FORMAT(3E10.3)
1014 FORMAT(///,5X,*MATERIAL PROPERTIES*,//,5X,*YOUNGS MODULUS FOR *, *
    1MATERIAL  =*,E10.3,/,5X,*POISSONS RATIO FOR MATERIAL  =*,
    1E10.3,/,5X,*CONSTANT ELEMENT THICKNESS   =*,E10.3,
    1//,5X,*DATA COMPLETE*)
1017 FORMAT(1H1,5X,*NODAL REACTIONS*,///,
    1          6X,4HNODE,6X,6HX COMP,6X,6HY COMP,/)
1018 FORMAT(1H1,///,5X,*NODAL DISPLACEMENTS*,///,
    1            19X,6HX COMP,9X,6HY COMP,//)
1019 FORMAT(5X,I5,5X,F10.5,5X,F10.5)
1020 FORMAT(1H1,///,5X,*ELEMENT STRESSES*,///,21X,*SIGMA X-X*,11X,
    1*SIGMA Y-Y*,13X,*TAU X-Y*,/)
1021 FORMAT(5X,I5,5X,3(F15.5,5X),/)
1025 FORMAT(I10,2(2X,E10.3))


    STOP
    END
```

Complete listing of the program with the band solver implemented

```
      PROGRAM SIMPLE(INPUT,OUTPUT,TAPE5=INPUT,TAPE6=OUTPUT)


      DIMENSION LNODS(65,3),COORDS(5),2),ALOAD(50,2),NFIX(50,2)
      DIMENSION ELSTIF(6,6),SS(90,90)
      DIMENSION SL(90),D(3,3),B(3,6),DB(3,6),DISPL(90),DISP(6)
      DIMENSION STRESS(65,3)
      DIMENSION REACT(2)
      INTEGER HBW


C     NVABZ IS THE NUMBER OF VARIABLES PER NODE
C     NNODZ IS THE NUMBER OF NODES PER ELEMENT


      HBW=0
      NVABZ=2
      NNODZ=3



C     READ IN DATA

      READ(5,1001) MAXNEL,MAXNOD,MAXLOD,MAXFIX
      WRITE(6,1002) MAXNEL,MAXNOD,MAXLOD,MAXFIX
```

74

```
C       INITIALIZE VECTORS AND MATRICES

        MAXVAR=MAXNOD*NVABZ
        DO 2 I=1,MAXNOD
        DO 1 J=1,2
        ALOAD(I,J)=0.0
        NFIX(I,J)=0.0
    1   CONTINUE
    2   CONTINUE
        DO 4 I=1,MAXVAR
        SL(I)=0.0
        DISPL(I)=0.0
        DO 3 J=1,MAXVAR
        SS(I,J)=0.0
    3   CONTINUE
    4   CONTINUE
        DO 6 I=1,MAXNEL
        DO 5 J=1,3
        STRESS(I,J)=0.0
    5   CONTINUE
    6   CONTINUE

C       READ IN ELEMENT DEFINITIONS

        WRITE(6,1003)
        DO 11 NEL=1,MAXNEL
        READ(5,1004) (LNODS(NEL,I),I=1,NNODZ)
        WRITE(6,1005) NEL,(LNODS(NEL,I),I=1,NNODZ)
        NIC1=LNODS(NEL,1)
        NIC2=LNODS(NEL,2)
        NIC3=LNODS(NEL,3)
        IF(IABS(NIC1-NIC2).GT.HBW) HBW=IABS(NIC1-NIC2)
        IF(IABS(NIC2-NIC3).GT.HBW) HBW=IABS(NIC2-NIC3)
        IF(IABS(NIC3-NIC1).GT.HBW) HBW=IABS(NIC3-NIC1)
   11   CONTINUE
        HBW=(HBW+1)*NVABZ

C       READ IN NODAL CO-ORDINATES

        WRITE(6,1006)
        DO 12 NOD=1,MAXNOD
        READ(5,1007) NIC,(COORDS(NIC,I),I=1,NVABZ)
        WRITE(6,1008) NIC,(COORDS(NIC,I),I=1,NVABZ)
   12   CONTINUE

C       READ IN APPLIED LOADS AND FIXITIES

        DO 14 J=1,MAXLOD
        READ(5,1010) NIC,(ALOAD(NIC,I),I=1,NVABZ)
   14   CONTINUE
        DO 15 J=1,MAXFIX
        READ(5,1011) NIC,(NFIX(NIC,I),I=1,NVABZ)
   15   CONTINUE

        WRITE(6,1009)
        DO 16 I=1,MAXNOD
        WRITE(6,1012) I,(ALOAD(I,J),J=1,NVABZ),(NFIX(I,J),J=1,NVABZ)
   16   CONTINUE

C       READ IN ELASTIC PROPERTIES

        READ(5,1013) YM,PR,T
        WRITE(6,1014) YM,PR,T
```

```
C      SET UP STRUCTURAL STIFFNESS MATRIX

       EC1=YM/(1-PR*PR)
       EC2=PR
       EC12=EC1*(1-EC2)/2.0


       DO 100 NEL=1,MAXNEL


C      SET UP ELEMENT GEOMETRY

       NIC1=LNODS(NEL,1)
       NIC2=LNODS(NEL,2)
       NIC3=LNODS(NEL,3)

       X1=COORDS(NIC1,1)
       Y1=COORDS(NIC1,2)
       X2=COORDS(NIC2,1)
       Y2=COORDS(NIC2,2)
       X3=COORDS(NIC3,1)
       Y3=COORDS(NIC3,2)

       AREA=(X2*Y3-X3*Y2-X1*(Y3-Y2)+Y1*(X3-X2))/2.0

       BI=Y2-Y3
       CI=X3-X2
       BJ=Y3-Y1
       CJ=X1-X3
       BM=Y1-Y2
       CM=X2-X1


C      SET UP ELEMENT STIFFNESS MATRIX

       ELSTIF(1,1)=EC1*BI**2.0+EC12*CI**2.0
       ELSTIF(2,1)=(EC1*EC2+EC12)*BI*CI
       ELSTIF(2,2)=EC1*CI**2.0+EC12*BI**2.0
       ELSTIF(3,1)=EC1*BI*BJ+EC12*CI*CJ
       ELSTIF(3,2)=EC1*EC2*BJ*CI+EC12*BI*CJ
       ELSTIF(3,3)=EC1*BJ**2.0+EC12*CJ**2.0
       ELSTIF(4,1)=EC1*EC2*BI*CJ+EC12*BJ*CI
       ELSTIF(4,2)=EC1*CI*CJ+EC12*BI*BJ
       ELSTIF(4,3)=(EC1*EC2+EC12)*BJ*CJ
       ELSTIF(4,4)=EC1*CJ**2.0+EC12*BJ**2.0
       ELSTIF(5,1)=EC1*BI*BM+EC12*CI*CM
       ELSTIF(5,2)=EC1*EC2*BM*CI+EC12*BI*CM
       ELSTIF(5,3)=EC1*BJ*BM+EC12*CJ*CM
       ELSTIF(5,4)=EC1*EC2*BM*CJ+EC12*BJ*CM
       ELSTIF(5,5)=EC1*BM**2.0+EC12*CM**2.0
       ELSTIF(6,1)=EC1*EC2*BI*CM+EC12*BM*CI
       ELSTIF(6,2)=EC1*CI*CM+EC12*BI*BM
       ELSTIF(6,3)=EC1*EC2*BJ*CM+EC12*BM*CJ
       ELSTIF(6,4)=EC1*CJ*CM+EC12*BJ*BM
       ELSTIF(6,5)=(EC1*EC2+EC12)*BM*CM
       ELSTIF(6,6)=EC1*CM**2.0+EC12*BM**2.0

       DO 40 I=1,5
       IP1=I+1
       DO 39 J=IP1,6
       ELSTIF(I,J)=ELSTIF(J,I)
    39 CONTINUE
    40 CONTINUE
```

76

```
       DO 42 I=1,6
       DO 41 J=1,6
       ELSTIF(I,J)=ELSTIF(I,J)*T/(4.0*AREA)
    41 CONTINUE
    42 CONTINUE

C      ADD IN ELEMENT STIFFNESS INTO STRUCTURAL MATRIX
```

3 nodes per element

```
       DO 50 I=1,NNODZ
       DO 49 II=1,NVABZ
```
2 variables per nodes.
```
       ISTRST=(LNODS(NEL,I)-1)*NVABZ+II
       IELEMT=(I-1)*NVABZ+II
       DO 48 J=1,NNODZ
       DO 47 JJ=1,NVABZ
       JSTRST=(LNODS(NEL,J)-1)*NVABZ+JJ
       JELEMT=(J-1)*NVABZ+JJ
       IF(ISTRST.GT.JSTRST) GO TO 47
       JSTRST=JSTRST-ISTRST+1
       SS(ISTRST,JSTRST)=SS(ISTRST,JSTRST)+ELSTIF(IELEMT,JELEMT)
    47 CONTINUE
    48 CONTINUE
    49 CONTINUE
    50 CONTINUE
   100 CONTINUE


C      ADD IN APPLIED LOADS


       DO 181 I=1,MAXNOD
       DO 180 J=1,NVABZ
       SL(2*I-2+J)=ALOAD(I,J)+SL(2*I-2+J)
   180 CONTINUE
   181 CONTINUE




C      APPLY FIXITIES


       I=0
       DO 200 NIC=1,MAXNOD
       DO 199 NVB=1,NVABZ
       I=I+1
       IF(NFIX(NIC,NVB).EQ.0) GO TO 199
       SS(I,1)=SS(I,1)+1.0E+50
   199 CONTINUE
   200 CONTINUE




C      CARRY OUT GAUSSIAN REDUCTION


       M1=MAXNOD*NVABZ
       M2=M1-1
       DO 300 I=1,M2
       I1=I+1
       I2=I+HBW-1
```

```
      IF(I2.GT.M1) I2=M1
      DO 290 K=I1,I2
      FACT=SS(I,K-I+1)/SS(I,1)
      NSF1=HBW-K+I
      DO 280 J=1,NSF1
      SS(K,J)=SS(K,J)-FACT*SS(I,J+K-I)
  280 CONTINUE
      SL(K)=SL(K)-FACT*SL(I)
  290 CONTINUE
  300 CONTINUE


C     BACK SUBSTITUTE

      DO 400 I=1,M1
      II=M1-I+1
      PIVOT=SS(II,1)
      SS(II,1)=0.0
      DO 350 J=1,HBW
      K=II+J-1
      IF(K.GT.M1) GO TO 350
      SL(II)=SL(II)-SS(II,J)*DISPL(K)
  350 CONTINUE
      DISPL(II)=SL(II)/PIVOT
  400 CONTINUE

      WRITE(6,1018)
      DO 401 I=1,MAXNOD
      WRITE(6,1019) I,DISPL(2*I-1),DISPL(2*I)
  401 CONTINUE

      WRITE(6,1017)
      DO 450 I=1,MAXNOD
      II=0
      DO 440 J=1,NVABZ
      REACT(J)=0.0
      IF(NFIX(I,J).EQ.0) GO TO 440
      III=(I-1)*NVABZ+J
      REACT(J)=-DISPL(III)*1.0E50
      II=1
  440 CONTINUE
      IF(II.EQ.1) WRITE(6,1025) I,REACT
  450 CONTINUE

C     CALAULATE STRESSES

      WRITE(6,1020)


      DO 602 I=1,3
      DO 601 J=1,3
      D(I,J)=0.0
  601 CONTINUE
  602 CONTINUE

      D(1,1)=EC1
      D(1,2)=EC1*EC2
      D(2,1)=EC1*EC2
      D(2,2)=EC1
      D(3,3)=EC12

      DO 700 NEL=1,MAXNEL

      DO 604 I=1,3
      DO 603 J=1,6
      B(I,J)=0.0
  603 CONTINUE
  604 CONTINUE
```

```
      NIC1=LNODS(NEL,1)
      NIC2=LNODS(NEL,2)
      NIC3=LNODS(NEL,3)

      X1=COORDS(NIC1,1)
      Y1=COORDS(NIC1,2)
      X2=COORDS(NIC2,1)
      Y2=COORDS(NIC2,2)
      X3=COORDS(NIC3,1)
      Y3=COORDS(NIC3,2)

      A=(X2*Y3-X3*Y2-X1*(Y3-Y2)+Y1*(X3-X2))/2.0

      B(1,1)=(Y2-Y3)/(2.0*A)
      B(1,3)=(Y3-Y1)/(2.0*A)
      B(2,4)=(X1-X3)/(2.0*A)
      B(2,2)=(X3-X2)/(2.0*A)
      B(1,5)=(Y1-Y2)/(2.0*A)
      B(2,6)=(X2-X1)/(2.0*A)
      DO 605 I=1,3
      B(3,2*I-1)=B(2,2*I)
      B(3,2*I)=B(1,2*I-1)
  605 CONTINUE

      DO 608 I=1,3
      DO 607 J=1,6
      DB(I,J)=0.0
      DO 606 K=1,3
      DB(I,J)=DB(I,J)+D(I,K)*B(K,J)
  606 CONTINUE
  607 CONTINUE
  608 CONTINUE

      DO 609 I=1,3
      II=2*LNODS(NEL,I)-1
      JJ=2*LNODS(NEL,I)
      DISP(2*I-1)=DISPL(II)
      DISP(2*I)=DISPL(JJ)
  609 CONTINUE

      DO 620 I=1,3
      DO 619 J=1,6
      STRESS(NEL,I)=STRESS(NEL,I)+DB(I,J)*DISP(J)
  619 CONTINUE
  620 CONTINUE
      WRITE(6,1021) NEL,(STRESS(NEL,J),J=1,3)
  700 CONTINUE

 1001 FORMAT(4I5)
 1002 FORMAT(1H1,5X,
     1*TOTAL NUMBER OF ELEMENTS =*,I5,/,25X,*NODES =*,I5,
     1/,25X,*LOADS =*,I5,/,22X,*FIXITIES =*,I5)
 1003 FORMAT(///,5X,*ELEMENT DEFINITIONS*,/,5X,*ELEMENT*,2X,*NODE1*,2X,
     1*NODE2*,2X,*NODE3*,/)
 1004 FORMAT(3I5)
 1005 FORMAT(5X,I5,3(2X,I5))
 1006 FORMAT(///,5X,*NODAL CO-ORDINATES*,/,5X,*NODE*,6X,*X*,9X,*Y*,/)
 1007 FORMAT(I5,2(5X,F10.6))
 1008 FORMAT(5X,I4,2F10.3)
 1010 FORMAT(I5,2(5X,F10.3))
 1011 FORMAT(3I5)
 1009 FORMAT(///,5X,*APPLIED LOADS AND FIXITIES*,/,5X,*NODE*,5X,
     1*APPLIED LOADS*,8X,*FIXITIES*,/)
 1012 FORMAT(5X,I4,3X,2E10.3,2I5)
```

```
1013 FORMAT(3E10.3)
1014 FORMAT(///,5X,*MATERIAL PROPERTIES*,//,5X,*YOUNGS MODULUS FOR *,
     1MATERIAL   =*,E10.3,/,5X,*POISSONS RATIO FOR MATERIAL   =*,
     1E10.3,/,5X,*CONSTANT ELEMENT THICKNESS   =*,E10.3,
     1//,5X,*DATA COMPLETE*)
1017 FORMAT(1H1,5X,*NODAL REACTIONS*,///,
     1        6X,4HNODE,6X,6HX COMP,6X,6HY COMP,/)
1018 FORMAT(1H1,///,5X,*NODAL DISPLACEMENTS*,///,
     1          19X,6HX COMP,9X,6HY COMP,//)
1019 FORMAT(5X,I5,5X,F10.5,5X,F10.5)
1020 FORMAT(1H1,///,5X,*ELEMENT STRESSES*,///,21X,*SIGMA X-X*,11X,
     1*SIGMA Y-Y*,13X,*TAU X-Y*,/)
1021 FORMAT(5X,I5,5X,3(F15.5,5X),/)
1025 FORMAT(I10,2(2X,E10.3))


     STOP
     END
```

References

1. M.J. Turner, R.W. Clough, H.C. Martin and L.J. Topp. Stiffness and deflection analysis of complex structures. *Journal of Aeronautical Science*, 23, 805-823, 1956.
2. E.P. Popov. *Mechanics of Materials*, 2nd ed. Englewood Cliffs, N.J., Prentice-Hall, 1976.
3. S.H. Crandall. *Engineering Analysis*. New York, McGraw-Hill Book Co., 1956.
4. V.N. Faddeeva. *Computational Methods of Linear Algebra*. New York, Dover Publications, 1956.

3 Quadratic Isoparametric Element for Plane Elasticity

3.1 Introduction

The constant strain element presented in Chapter 2 is the lowest order element available, and an element with somewhat better characteristics is a quadrilateral element with four nodes. At first sight it seems possible to apply the displacement functions of a rectangular element (Table 1.1) directly to the quadrilateral element, such that

$$u = A_1 + A_2 x + A_3 y + A_4 xy$$
$$v = B_1 + B_2 x + B_3 y + B_4 xy$$
(3.1a)

However, upon closer examination it will be found that this is unsuitable since the displacements will vary parabolically along the edges which are in general not parallel to the x and y axes. With only two nodes along an edge it is no longer possible to guarantee the compatibility of displacements between adjacent elements.

Eqn.(3.1a) can be rewritten in non-dimensional form as

$$u = A_1 + A_2 x/a + A_3 y/b + A_4 xy/ab$$

or

$$u = A_1 + A_2 \xi + A_3 \eta + A_4 \xi\eta$$
$$v = B_1 + B_2 \xi + B_3 \eta + B_4 \xi\eta$$
(3.1b)

The coordinates in ξ and η in Eqn.(3.1b) vary between -1 and $+1$, as shown in Fig.3.1(a). It can be seen that it is possible to use another set of curvilinear coordinates for quadrilateral element such that ξ and η, while not simply equal to x/a and y/b respectively as in Eqn. (3.1b) and is to be explained in detail subsequently, will nevertheless take up unit values at the inclined edges (Fig.3.1(b)). It should be noted that the ξ and η axes will bisect the edges in each case.

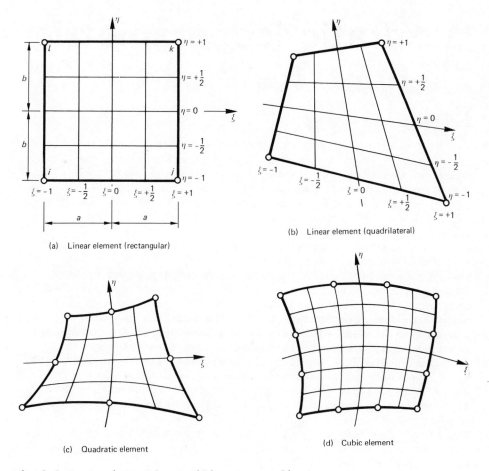

(a) Linear element (rectangular)

(b) Linear element (quadrilateral)

(c) Quadratic element

(d) Cubic element

Fig.3.1 Cartesian and curvilinear coordinates

3.2 Isoparametric concept [1]

The displacement field for a two-dimensional element of general shape is given by

$$u = N_1u_1 + N_2u_2 + N_3u_3 + \ldots + N_ru_r$$
$$= \sum N_iu_i$$

$$v = N_1v_1 + N_2v_2 + N_3v_3 + \ldots + N_rv_r$$
$$= \sum N_iv_i$$

(3.2)

in which N_i is a function of the curvilinear coordinates ξ and η, and r is the number of nodes.

The coordinates x and y inside the element domain can be described in a similar way by

$$x = M_1x_1 + M_2x_2 + M_3x_3 + \ldots + M_rx_r$$

$$= \sum M_i x_i$$

(3.3)

$$y = M_1y_1 + M_2y_2 + M_3y_3 + \ldots + M_ry_r$$

$$= \sum M_i y_i$$

in which M_i is also a function of ξ and η.

For the particular case in which N_i and M_i are identical, i.e. the shape functions defining the displacement fields and geometry are the same, the element is termed isoparametric. Eqn.(3.3) provides the relationship between the Cartesian and the curvilinear coordinate systems, so that appropriate transformations can be carried out in the stiffness formulation.

There are a number of interesting points concerning the isoparametric concept and they are discussed below.

(i) The isoparametric concept allows any arbitrary geometry to be closely approximated, thereby minimizing any error associated with modelling the geometry and without resorting to the use of a fine mesh along the boundaries. This is because the geometry of the edges of an element will vary in the same way as the displacement function, i.e. a linear element will have straight edges all round, a quadratic element will have edges varying as parabolas, a cubic element will have edges varying as a third-degree polynomial curve, etc., as shown in Fig.3.1. For curved boundaries it is thus more desirable to use a higher order isoparametric element.

(ii) The rigid body displacement criterion discussed in Section 1.8 is satisfied for isoparametric elements. As we can see, Eqn.(1.28b) and (1.28c) are satisfied automatically since

$$x = \sum N_i x_i \quad \text{and} \quad y = \sum N_i y_i$$

by definition, while Eqn.(1.28a) requires that $\sum N_i = 1$ is also

valid for all isoparametric elements because the shape functions are symmetric with respect to the coordinates ξ, η, and will cancel out during the summation process. Incidentally, this symmetry may also be an advantage from the programming point of view, and for the integration of the stiffness coefficients only a quarter of the function and derivative values need to be calculated while the rest are obtained by simple cyclic permutation.

(iii) The constant strain criterion is satisfied. This can be proved by proceeding as described in Section 1.8 and assigning a 'constant strain' field involving only constants and linear variables as given below.

$$f = A_1 + A_2x + A_3y$$

(3.4)

At any node i the following equality exists,

$$f_i = A_1 + A_2x_i + A_3y_i$$

(3.5a)

and, substituting into Eqn.(3.2),

$$f = A_1 \sum N_i + A_2 \sum N_i x_i + A_3 \sum N_i y_i \qquad (3.5b)$$

By comparing Eqn.(3.5b) with Eqn.(3.4), it is seen that the constant strain criterion is satisfied, provided that

$$\sum N_i = 1 \qquad (3.6a)$$

$$\sum N_i x_i = x \qquad (3.6b)$$

$$\sum N_i y_i = y \qquad (3.6c)$$

Eqn.(3.6) is in fact a restatement of the rigid body displacement criterion discussed above and in Section 1.8.
(iv) Unlike the Lagrange elements (*see* Section 1.7), the commonly used isoparametric elements (up to the cubic element) have no internal nodes and are therefore more efficient from the computational point of view.

3.3 Family of isoparametric element

A whole family of two-dimensional isoparametric elements can be formulated but only three of them are shown in Fig.3.1, and of these three, the quadratic element is considered to be the best and will be used in the example program. Elements of an order higher than three are not really practical owing to the excessive computer time required for the generation of stiffness matrices.

At this stage it is worthwhile for us to re-examine the Lagrange polynomials given in Table 1.2. Obviously they can be utilized in the construction of shape functions for isoparametric elements although it is necessary to shift the coordinate origin from the left end to the centre and to change the variables from x/ℓ to $(1 + \xi)/2$ in the expressions. The modified results are given in Table 3.1.

3.3.1 LINEAR ELEMENT (Fig.3.1(b))

This element is simply a more general form of the linear Lagrange element given in Fig.1.6(a). The displacement functions in simple polynomial form have been given in Eqn.(3.1b) while the shape functions can be constructed directly as products of the linear Lagrange polynomials given in Table 3.1, and are of the form

$$N = \mathcal{L}^1(\xi)\, \mathcal{L}^1(\eta)$$

More specifically, we have

at node i, $\xi_i = -1$, $\eta_i = -1$ and $N_i = \frac{1}{4}(1 - \xi)(1 - \eta)$

at node j, $\xi_j = +1$, $\eta_j = -1$ and $N_j = \frac{1}{4}(1 + \xi)(1 - \eta)$

at node k, $\xi_k = +1$, $\eta_k = +1$ and $N_k = \frac{1}{4}(1 + \xi)(1 + \eta)$

at node ℓ, $\xi_\ell = -1$, $\eta_\ell = +1$ and $N_\ell = \frac{1}{4}(1 - \xi)(1 + \eta)$

TABLE 3.1 *Lagrange polynomials in dimensionless form*

$$\mathcal{L}_0^1(\xi) = \frac{1}{2}(1 - \xi)$$

$$\mathcal{L}_1^1(\xi) = \frac{1}{2}(1 + \xi)$$

$$\mathcal{L}_0^2(\xi) = \frac{1}{2}\xi(\xi - 1)$$

$$\mathcal{L}_1^2(\xi) = (1 - \xi^2)$$

$$\mathcal{L}_2^2(\xi) = \frac{1}{2}(1 + \xi)\xi$$

$$\mathcal{L}_0^3(\xi) = \frac{1}{16}(3\xi + 1)(3\xi - 1)(1 - \xi)$$

$$\mathcal{L}_1^3(\xi) = \frac{9}{16}(1 - \xi^2)(1 - 3\xi)$$

$$\mathcal{L}_2^3(\xi) = \frac{9}{16}(1 - \xi^2)(1 + 3\xi)$$

$$\mathcal{L}_3^3(\xi) = \frac{1}{16}(1 + \xi)(3\xi + 1)(3\xi - 1)$$

or, using i as the general subscript for all nodes,

$$N_i = \frac{1}{4}(1 + \xi\xi_i)(1 + \eta\eta_i) \tag{3.7}$$

in which ξ_i and η_i are the coordinate values of node i.

3.3.2 QUADRATIC ELEMENT (Fig.3.1(c))

There are altogether eight nodes for this element and with only three nodes along one edge, the displacement variation should be parabolic to satisfy compatibility. The correct displacement functions are

$$u = A_1 + A_2\xi + A_3\eta + A_4\xi^2 + A_5\xi\eta + A_6\eta^2 + A_7\xi^2\eta + A_8\xi\eta^2$$

$$v = B_1 + B_2\xi + B_3\eta + B_4\xi^2 + B_5\xi\eta + B_6\eta^2 + B_7\xi^2\eta + B_8\xi\eta^2 \tag{3.8}$$

As far as shape functions are concerned, for corner nodes they should vary as parabolas in the ξ and η directions and by definition, they should always have zero values at the midside nodes (Fig.3.2(c)). For

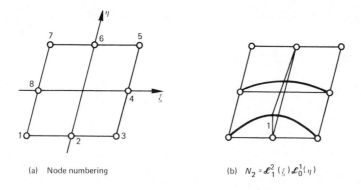

(a)   Node numbering

(b) $N_2 = \mathscr{L}_1^2(\xi)\mathscr{L}_0^1(\eta)$

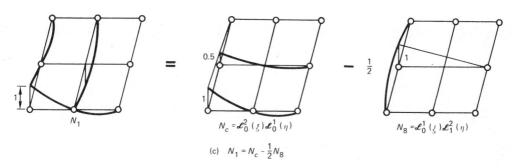

$N_c = \mathscr{L}_0^2(\xi)\mathscr{L}_0^1(\eta)$

$N_8 = \mathscr{L}_0^1(\xi)\mathscr{L}_1^2(\eta)$

(c) $N_1 = N_c - \frac{1}{2}N_8$

Fig.3.2 Construction of quadratic shape functions

midside nodes the construction is fairly straightforward and using node 2 in Fig.3.2(a) as an example we have

$$N_2 = \mathscr{L}_1^2(\xi)\mathscr{L}_0^1(\eta)$$
$$= \frac{1}{2}(1 - \xi)^2(1 - \eta)$$

In general for midside nodes with $\xi_i = 0$,

$$N_i = \frac{1}{2}(1 - \xi^2)(1 + \eta\eta_i) \qquad (3.9a)$$

and for midside nodes with $\eta_i = 0$

$$N_i = \frac{1}{2}(1 + \xi\xi_i)(1 - \eta^2) \qquad (3.9b)$$

The construction of the shape function for a corner node is more complex. Referring to Fig.3.2(c) it is seen that a shape function for node 1

$$N_c = \mathscr{L}_0^2(\xi)\mathscr{L}_0^1(\eta)$$

will satisfy the requirements along the edges $\eta = \pm 1$, and $\xi = +1$, although at node 8 on the edge $\xi = -1$, the function is equal to 0.5 instead of zero. This value can be corrected by subtracting Eqn.(3.9b) from N_c in the manner shown in Fig.3.2(c), i.e.

$$N_1 = N_c - \frac{1}{2}N_8$$

$$= \mathcal{L}_0^2(\xi)\,\mathcal{L}_0^1(\eta) - \frac{1}{2}\mathcal{L}_0^1(\xi)\,\mathcal{L}_1^2(\eta)$$

$$= \frac{1}{4}\xi(\xi - 1)(1 - \eta) - \frac{1}{4}(1 - \xi)(1 - \eta^2)$$

$$= \frac{1}{4}(1 - \xi)(1 - \eta)(-\xi - 1 - \eta)$$

The general equation for shape functions at all corner nodes is

$$N_i = \frac{1}{4}(1 + \xi\xi_i)(1 + \eta\eta_i)(\xi\xi_i + \eta\eta_i - 1) \qquad (3.9c)$$

Note that it is also possible to construct the shape function by starting with

$$N_c = \mathcal{L}_0^1(\xi)\,\mathcal{L}_0^1(\eta)$$

and then writing

$$N_1 = N_c - \frac{1}{2}N_2 - \frac{1}{2}N_8$$

3.3.3 CUBIC ELEMENT (Fig.3.1(d))

The displacement functions in simple polynomial form are

$$u = A_1 + A_2\xi + A_3\eta + A_4\xi^2 + A_5\xi\eta + A_6\eta^2 + A_7\xi^3 + A_8\xi^2\eta$$

$$+ A_9\xi\eta^2 + A_{10}\eta^3 + A_{11}\xi^3\eta + A_{12}\xi\eta^3$$

$$(3.10)$$

$$v = B_1 + B_2\xi + B_3\eta + B_4\xi^2 + B_5\xi\eta + B_6\eta^2 + B_7\xi^3 + B_8\xi^2\eta$$

$$+ B_9\xi\eta^2 + B_{10}\eta^3 + B_{11}\xi^3\eta + B_{12}\xi\eta^3$$

which conform to the requirements of having twelve nodes and cubic variations of displacements along the edges.

The construction of the shape functions along the same line as that indicated for the quadratic element and as an example, for node 2 with $\xi = -1$ and $\eta = -\frac{1}{3}$,

$$N_2 = \mathcal{L}_1^3(\xi)\,\mathcal{L}_0^1(\eta)$$

$$= \frac{9}{32}(1 - \xi^2)(1 - \eta)(1 - 3\xi)$$

or in general, for all edge nodes with $\xi_i = \pm \frac{1}{3}$,

$$N_i = \frac{9}{32}(1 - \xi^2)(1 + \eta\eta_i)(1 + 9\xi\xi_i) \qquad (3.11a)$$

For node 12 with $\xi = -1$ and $\eta = -\frac{1}{3}$

$$N_{12} = \mathcal{L}_0^1(\xi)\,\mathcal{L}_1^3(\eta)$$

$$= \frac{9}{32}(1 - \xi)(1 - \eta^2)(1 - 3\eta)$$

or in general, for all edge nodes with $\eta_i = \pm \frac{1}{3}$,

$$N_i = \frac{9}{32}(1 + \xi\xi_i)(1 - \eta^2)(1 + 9\eta\eta_i)$$ (3.11b)

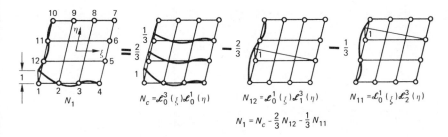

$$N_c = \mathcal{L}_0^3(\xi)\,\mathcal{L}_0^1(\eta) \qquad N_{12} = \mathcal{L}_0^1(\xi)\,\mathcal{L}_1^3(\eta) \qquad N_{11} = \mathcal{L}_0^1(\xi)\,\mathcal{L}_2^3(\eta)$$

$$N_1 = N_c - \frac{2}{3}N_{12} - \frac{1}{3}N_{11}$$

Fig.3.3 *Construction of a cubic shape function for corner nodes*

For corner nodes, the same procedure of correction as previously demonstrated for the quadratic element has to be made. From Fig.3.3 it is seen that a shape function N_c for node 1 will yield incorrect values at nodes 11 and 12, which should be corrected by the values $\frac{1}{3}$ and $\frac{2}{3}$ respectively. Thus the proper shape function for node 1 is

$$N_1 = N_c - \frac{2}{3}N_{12} - \frac{1}{3}N_{11}$$

$$= \mathcal{L}_0^3(\xi)\,\mathcal{L}_0^1(\eta) - \frac{2}{3}\mathcal{L}_0^1(\xi)\,\mathcal{L}_1^3(\eta) - \frac{1}{3}\mathcal{L}_0^1(\xi)\,\mathcal{L}_2^3(\eta)$$

$$= \frac{1}{32}(3\xi + 1)(3\xi - 1)(1 - \xi)(1 - \eta)$$

$$- \frac{6}{32}(1 - \xi)(1 - \eta^2)(1 - 3\eta) - \frac{3}{32}(1 - \xi)(1 - \eta^2)(1 + 3\eta)$$

$$= \frac{1}{32}(1 - \xi)(1 - \eta)\left[-10 + 9(\xi^2 + \eta^2)\right]$$

The general shape function for all corner nodes is

$$N_i = \frac{1}{32}(1 + \xi\xi_i)(1 + \eta\eta_i)\left[-10 + 9(\xi^2 + \eta^2)\right]$$ (3.11c)

3.4 Stiffness matrix formulation

It has already been shown in Chapter 1 that the stiffness matrix of an element can be derived through either a total potential energy or a virtual work approach. The expression (1.46) for element stiffness has been used a number of times already and is of the general form

$$[k] = \int [B]^T [D][B]\,d(\text{vol})$$

For two-dimensional problems,

$$[k] = t \int [B]^T [D][B] \mathrm{d(area)} \qquad (3.12)$$

The $[B]$ matrix given in Eqn. (3.12) expresses strains ε_x, ε_y, γ_{xy} in terms of the nodal displacements and therefore contains first derivatives of the shape functions with respect to the global axes x and y. For the quadratic element the strain matrix $[B]$ is

$$[B] = \begin{bmatrix} \dfrac{\partial N_1}{\partial x} & 0 & \dfrac{\partial N_2}{\partial x} & 0 & \cdots & \dfrac{\partial N_8}{\partial x} & 0 \\[2ex] 0 & \dfrac{\partial N_1}{\partial y} & 0 & \dfrac{\partial N_2}{\partial y} & \cdots & 0 & \dfrac{\partial N_8}{\partial y} \\[2ex] \dfrac{\partial N_1}{\partial y} & \dfrac{\partial N_1}{\partial x} & \dfrac{\partial N_2}{\partial y} & \dfrac{\partial N_2}{\partial x} & \cdots & \dfrac{\partial N_8}{\partial y} & \dfrac{\partial N_8}{\partial x} \end{bmatrix}$$

and a typical submatrix is given by

$$[B_i] = \begin{bmatrix} \dfrac{\partial N_i}{\partial x} & 0 \\[2ex] 0 & \dfrac{\partial N_i}{\partial y} \\[2ex] \dfrac{\partial N_i}{\partial y} & \dfrac{\partial N_i}{\partial x} \end{bmatrix} \qquad (3.13)$$

However, the shape functions for isoparametric elements (*see* Section 3.3) are defined with respect to the curvilinear coordinates ξ and η and therefore they cannot be differentiated directly with respect to the global x, y axes.

In order to overcome this difficulty it is necessary to obtain a relationship between the derivatives of the two sets of coordinates and this can be achieved through the normal chain rule of partial differentiation. For a two-dimensional problem the derivatives are related by

$$\left. \begin{aligned} \frac{\partial N}{\partial \xi} &= \frac{\partial N}{\partial x}\frac{\partial x}{\partial \xi} + \frac{\partial N}{\partial y}\frac{\partial y}{\partial \xi} \\[2ex] \frac{\partial N}{\partial \eta} &= \frac{\partial N}{\partial x}\frac{\partial x}{\partial \eta} + \frac{\partial N}{\partial y}\frac{\partial y}{\partial \eta} \end{aligned} \right\} \qquad (3.14)$$

which in matrix form gives

$$\left\{ \begin{aligned} \frac{\partial N}{\partial \xi} \\[2ex] \frac{\partial N}{\partial \eta} \end{aligned} \right\} = \begin{bmatrix} \dfrac{\partial x}{\partial \xi} & \dfrac{\partial y}{\partial \xi} \\[2ex] \dfrac{\partial x}{\partial \eta} & \dfrac{\partial y}{\partial \eta} \end{bmatrix} \left\{ \begin{aligned} \frac{\partial N}{\partial x} \\[2ex] \frac{\partial N}{\partial y} \end{aligned} \right\} = [J] \left\{ \begin{aligned} \frac{\partial N}{\partial x} \\[2ex] \frac{\partial N}{\partial y} \end{aligned} \right\} \qquad (3.15)$$

The matrix $[J]$ relating the derivatives of the two systems is called the Jacobian matrix and its coefficients can be obtained by differentiating Eqn.(3.3) with respect to the ξ and η coordinates.

For the quadratic element,

$$[J] = \begin{bmatrix} \dfrac{\partial N_1}{\partial \xi} & \dfrac{\partial N_2}{\partial \xi} & \cdots & \dfrac{\partial N_8}{\partial \xi} \\[2ex] \dfrac{\partial N_1}{\partial \eta} & \dfrac{\partial N_2}{\partial \eta} & \cdots & \dfrac{\partial N_8}{\partial \eta} \end{bmatrix} \begin{bmatrix} x_1 & y_1 \\ x_2 & y_2 \\ \cdot & \cdot \\ \cdot & \cdot \\ \cdot & \cdot \\ x & y \end{bmatrix} \tag{3.16}$$

$$= \begin{bmatrix} \dfrac{\partial \sum N_i x_i}{\partial \xi} & \dfrac{\partial \sum N_i y_i}{\partial \xi} \\[2ex] \dfrac{\partial \sum N_i x_i}{\partial \eta} & \dfrac{\partial \sum N_i y_i}{\partial \eta} \end{bmatrix}$$

With $[J]$ determined it is a straightforward matter to express $\partial N/\partial x$, $\partial N/\partial y$ with respect to $\partial N/\partial \xi$, $\partial N/\partial \eta$ through the inverse of $[J]$.

$$\begin{Bmatrix} \dfrac{\partial N}{\partial x} \\[2ex] \dfrac{\partial N}{\partial y} \end{Bmatrix} = [J]^{-1} \begin{Bmatrix} \dfrac{\partial N}{\partial \xi} \\[2ex] \dfrac{\partial N}{\partial \eta} \end{Bmatrix} \tag{3.17}$$

To complete the transformation between the two systems it is also necessary to express the elemental area in Eqn.(3.12) in the form of $d\xi\,d\eta$.

Consider an elemental parallelogram enclosed by two vectors

$$\vec{d\xi} = \begin{Bmatrix} \dfrac{\partial x}{\partial \xi} \\[2ex] \dfrac{\partial y}{\partial \xi} \end{Bmatrix} d\xi \quad \text{and} \quad \vec{d\eta} = \begin{Bmatrix} \dfrac{\partial x}{\partial \eta} \\[2ex] \dfrac{\partial y}{\partial \eta} \end{Bmatrix} d\eta \tag{3.18}$$

with the vectors defined from the relationship between the x,y and ξ,η coordinate systems and directed tangentially to the ξ = constant and η = constant contours respectively. The cross-product of the two vectors is equal to the area of the elemental parallelogram concerned and thus

$$d(area) = \vec{d\xi} \times \vec{d\eta}$$

$$= \det \begin{bmatrix} \dfrac{\partial x}{\partial \xi} & \dfrac{\partial x}{\partial \eta} \\[2mm] \dfrac{\partial y}{\partial \xi} & \dfrac{\partial y}{\partial \eta} \end{bmatrix} (d\xi \quad d\eta)$$

$$= \det[J]d\xi d\eta \tag{3.19}$$

It follows that the integration limits should be changed to ± 1 and Eqn.(3.12) can now be written as

$$[K] = t \int_{-1}^{+1}\int_{-1}^{+1} [B]^T[D][B]\det[J]d\xi d\eta \tag{3.20}$$

with the understanding that $[B]^T[D][B]$ is a function of ξ and η only.

A necessary requirement for working out Eqn.(3.17) to obtain $\partial N/\partial x$ and $\partial N/\partial y$ is that $[J]$ can be inverted. This inverse exists if there is no excessive distortion of the element such that lines of constant ξ or η intersect inside or on the element boundaries, as shown in Fig.3.4.

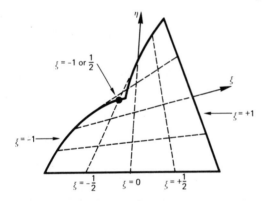

Fig.3.4 *Excessive distortion of an isoparametric element*

When the element is degenerated into a triangle by increasing an internal angle to 180° then $[J]$ is singular at that corner. It is, however, possible to obtain the element stiffness because $[J]$ is still unique at the Gauss points, but should the stresses be determined at that corner they will be indeterminate. A similar situation occurs when two adjacent corner nodes are made coincident to produce a triangular element. Therefore to ensure that $[J]$ can be inverted, any internal angle of each corner node of the element should be less than 180°, and, as an internal angle approaches 180° there is a loss in accuracy in the element stress, particularly at that corner.

In the majority of cases, the complexity of the expressions for the stiffness coefficients will make the explicit integration of Eqn.(3.20) exceedingly difficult, if not impossible, and thus numerical integration is mandatory for nearly all cases.

3.5 Numerical integration

For complex functions which cannot be integrated explicitly, a numerical integration can be carried out by first of all approximating the function with a polynomial. For example, if a linear approximation is made in place of the function shown in Fig.3.5(a), the integral, which represents the area under the curve, will be approximated by the area of a trapesium.

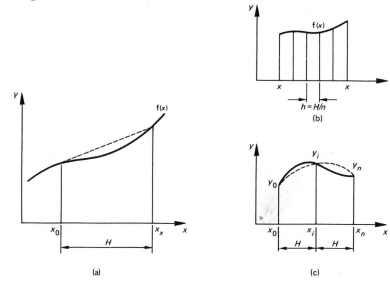

Fig.3.5 *Linear approximation of a curve*

$$I = \int_{x_0}^{x_n} f(x)\,dx \simeq \frac{f(x_0) + f(x_n)}{2}H$$

$$\simeq \frac{y_0 + y_n}{2}H \tag{3.21a}$$

Should the function depart appreciably from linearity then a significant error is likely to be introduced. This error can be reduced, however, by further subdividing the interval between x_0 and x_n into n divisions, as shown in Fig.3.5(b). Thus

$$I \simeq \frac{y_0 + y_1}{2}h + \frac{y_1 + y_2}{2}h + \dots + \frac{y_{n-1} + y_n}{2}h$$

$$\simeq \frac{h}{2}(y_0 + 2y_1 + 2y_2 + \dots + 2y_{n-1} + y_n) \tag{3.21b}$$

Eqn.(3.21) is the well-known trapezoidal rule, but it is rarely used for finite element work because of its inefficiency.

An alternative is to approximate the function by a higher-order polynomial, this time in the form of a parabola through three points, as shown in Fig.3.5(c).

The area under the parabola is given by

$$I \simeq \frac{H}{6}(y_0 + 4y_i + y_n) \tag{3.22a}$$

and if a further subdivision is made we have Simpson's rule and the integral is

$$I \simeq \frac{h}{6}\big[y_0 + y_n + 4(y_1 + y_3 + \ldots + y_{n-1})$$

$$+ 2(y_2 + y_4 + \ldots + y_{n-2})\big] \tag{3.22b}$$

Even higher-order polynomials can be used to approximate the function so as to improve the accuracy, and the general form of the integral is

$$I = \sum_{i=0}^{n} W_i f(x_i) \tag{3.23}$$

where $n + 1$ is the number of sample points for the polynomial. In Eqn. (3.23) the weighting coefficients are constant for equally spaced sampling points (as in Eqn.(3.22a)), and as a result for N points there are only N variables to fit a polynomial of order $N-1$. However, if unequally spaced sampling points are allowed and they are located so as to give the best approximation of the polynomial, there will be $2N$ variables to fit a polynomial of order $2N-1$. This is obviously a more efficient approach and is known as Gaussian quadrature[2].

For reasons of generality the coordinates of the sampling or Gauss points and the weighting coefficients are usually defined for limits of integration between -1 and +1, i.e.

$$\int_{-1}^{+1} f(\xi)\, d\xi = \sum_{i=1}^{N} W_i f(\xi_i) \tag{3.24}$$

which fits partly with the requirements of Eqn.(3.20). The values of W_i and ξ_i for various values of N are given in Table 3.2.

To extend Eqn.(3.24) to two dimensions so as to meet fully the requirement of Eqn.(3.20), it is required to integrate first with respect to one variable and then with respect to the other. Thus

$$\int_{-1}^{+1}\int_{-1}^{+1} f(\xi,\eta)\, d\xi d\eta = \int_{-1}^{+1}\left[\int_{-1}^{+1} f(\xi,\eta)\, d\xi\right] d\eta$$

$$= \int_{-1}^{1}\left[\sum_{i=1}^{N} W_i f(\xi_i,\eta)\right] d\eta$$

$$= \int_{-1}^{+1}\left[\sum_{i=1}^{N} W_i g_i(\eta)\right] d\eta$$

$$= \sum_{j=1}^{M}\sum_{i=1}^{N} W_i W_j g(\eta_j)$$

$$= \sum_{j=1}^{M}\sum_{i=1}^{N} W_i W_j f(\xi_i,\eta_j) \tag{3.25}$$

in which M and N can be equal or different.

TABLE 3.2 *Abscissae and weight coefficients of the Gaussian quadrature formula*

$$\int_{-1}^{1} f(x)\,dx = \sum_{i=1}^{n} W_i f(a_i)$$

	± a			W	
n = 2					
0.577 35	026 91	896 26	1.000 00	000 00	000 00
n = 3					
0.774 59	666 92	414 83	0.555 55	555 55	555 55
0.000 00	000 00	000 00	0.888 88	888 88	888 89
n = 4					
0.861 13	631 15	940 53	0.347 85	484 51	374 54
0.339 98	104 35	848 56	0.652 14	515 48	625 46
n = 5					
0.906 17	984 59	386 64	0.236 92	688 50	561 89
0.538 46	931 01	056 83	0.478 62	867 04	993 66
0.000 00	000 00	000 00	0.568 88	888 88	888 89
n = 6					
0.932 46	951 42	031 52	0.171 32	449 23	791 70
0.661 20	938 64	662 65	0.360 76	157 30	481 39
0.238 61	918 60	831 97	0.467 91	393 45	726 91
n = 7					
0.949 10	791 23	427 59	0.129 48	496 61	688 70
0.741 53	118 55	993 94	0.279 70	539 14	892 77
0.405 84	515 13	773 97	0.381 83	005 05	051 19
0.000 00	000 00	000 00	0.417 95	918 36	734 69
n = 8					
0.960 28	985 64	975 36	0.101 22	853 62	903 76
0.796 66	647 74	136 27	0.222 38	103 44	533 74
0.525 53	240 99	163 29	0.313 70	664 58	778 87
0.183 43	464 24	956 50	0.362 68	378 33	783 62
n = 9					
0.968 16	023 95	076 26	0.081 27	438 83	615 74
0.836 03	110 73	266 36	0.180 64	816 06	948 57
0.613 37	143 27	005 90	0.260 61	069 64	029 35
0.324 25	342 34	038 09	0.312 34	707 70	400 03
0.000 00	000 00	000 00	0.330 23	935 50	012 60
n = 10					
0.973 90	652 85	717 72	0.066 67	134 43	086 88
0.865 06	336 66	889 85	0.149 45	134 91	505 81
0.679 40	956 82	990 24	0.219 08	636 25	159 82
0.433 39	539 41	292 47	0.269 26	671 93	099 96
0.148 87	433 89	816 31	0.295 52	422 47	147 53

In the isoparámetric formulation of the element stiffness of a rectangular eight-nodal element the $[B]^T[D][B]$ to be integrated is a
fourth-order function in both directions. To integrate this exactly
requires more Gauss points than are in a 2 × 2 mesh but less than the
number of points present in a 3 × 3 mesh. Zienkiewicz [3] argues that
in the limit the element will be subjected to constant stress and hence
the integral required is the lowest order necessary to evaluate accur-
ately the area of the element, i.e. 3 × 3. On the other hand, the
displacement formulation of the finite element analysis yields a lower
bound on the strain energy and thereby overestimating the stiffness,
while by not integrating the coefficients exactly a lower stiffness is
obtained. Therefore by using a reduction in the integration order that
is required for exact integration a better solution is often obtained
because the two effects tend to cancel out each other. This is known as
the 'reduced integration technique'.

Bathe and Wilson [4] indicated that the best integrating order for a
rectangular eight-noded isoparametric element is 2 × 2. For an eight-
noded quadrilateral element $\det[J]$ is no longer a constant and has to be
included in the integration as an extra variable and therefore a higher-
order integration has to be used. For exact integration a 4 × 4 Gauss
point mesh should be used but better displacements will actually result
from using a 3 × 3 mesh due to reduced integration.

3.6 Calculation of element stiffness

In a typical finite element program using numerically integrated isoparametric elements and an advanced solution routine such as the front
solver, the computer time spent in calculating the element stiffness can
vary from 30 to 80 per cent of the total time required for a complete
analysis, and therefore the algorithm used to compute the element stiffness has a significant effect on the overall efficiency of the program.
With the algorithms in common use, the computer time required to calculate an element stiffness matrix varies by a factor of nine between
the best and the least efficient.

For the quadratic element, the stiffness matrix is of the form

$$[K] = \begin{bmatrix} k_{11} & k_{12} & k_{13} & k_{14} & k_{15} & k_{16} & k_{17} & k_{18} \\ & k_{22} & k_{23} & k_{24} & k_{25} & k_{26} & k_{27} & k_{28} \\ & & k_{33} & k_{34} & k_{35} & k_{36} & k_{37} & k_{38} \\ & & & k_{44} & k_{45} & k_{46} & k_{47} & k_{48} \\ & & & & k_{55} & k_{56} & k_{57} & k_{58} \\ & \text{symmetrical} & & & & k_{66} & k_{67} & k_{68} \\ & & & & & & k_{77} & k_{78} \\ & & & & & & & k_{88} \end{bmatrix} \qquad (3.26)$$

For efficiency of core storage, only the upper triangular matrix is used and the coefficients are stored in a one-dimensional array as shown below

$$
\begin{bmatrix}
k_{11} & k_{12} & k_{22} & | & k_{13} & k_{14} & k_{23} & k_{24} & | & k_{15} & \cdots & \cdots & k_{28} & | \\
k_{33} & k_{34} & k_{44} & | & k_{35} & \cdots & \cdots & k_{88} &
\end{bmatrix}
\tag{3.27}
$$

The broken lines in Eqn.(3.26) indicate a $[k]_{ij}$ submatrix which results from Eqns.(3.20) and (3.25) with the value of t in Eqn.(3.20) being taken to be unity.

$$
[k]_{ij} = \sum_{p=1}^{N}\sum_{q=1}^{N} W_p W_q
\begin{bmatrix}
\dfrac{\partial N_i}{\partial x} & O & \dfrac{\partial N_i}{\partial y} \\
O & \dfrac{\partial N_i}{\partial y} & \dfrac{\partial N_i}{\partial x}
\end{bmatrix}_{2\times 3}
\begin{bmatrix}
D_1 & D_2 & O \\
D_2 & D_4 & O \\
O & O & D_5
\end{bmatrix}_{3\times 3}
\begin{bmatrix}
\dfrac{\partial N_j}{\partial x} & O \\
O & \dfrac{\partial N_j}{\partial y} \\
\dfrac{\partial N_j}{\partial y} & \dfrac{\partial N_j}{\partial x}
\end{bmatrix}_{3\times 2}
\det[J]
$$

$$
= \begin{bmatrix}
k_1 & k_2 \\
k_3 & k_4
\end{bmatrix}_{ij}
\begin{pmatrix}
\dfrac{\partial N_i}{\partial x}D_1 & \dfrac{\partial N_i}{\partial x}D_2 & \dfrac{\partial N_i}{\partial y}D_5 \\
\dfrac{\partial N_i}{\partial y}D_2 & \dfrac{\partial N_i}{\partial y}D_4 & \dfrac{\partial N_i}{\partial x}D_5
\end{pmatrix}
\tag{3.28}
$$

Note that in Eqn.(3.28) only three coefficients need to be calculated when $i = j$ because of symmetry, and this corresponds to the storage arrangement given in Eqn.(3.27).

There are several methods of evaluating the expression in Eqn.(3.28). The most inefficient one is to carry out the full matrix multiplication of the triple product, and then to multiply it by the scalar quantity $W_p W_q \det[J]$ to give the required matrix, and finally to sum the matrices for all the Gauss points.

A far more efficient method is to multiply the matrices by hand to obtain explicit expressions for k_1ij, k_2ij, k_3ij and k_4ij. Due to the presence of a number of zero coefficients in both $[B]$ and $[D]$ matrices, the resulting expressions are relatively simple. Writing $W = W_p W_q \det[J]$ we have

$$
k_1ij = \sum\sum W\left(D_1\frac{\partial N_i}{\partial x}\frac{\partial N_j}{\partial x} + D_5\frac{\partial N_i}{\partial y}\frac{\partial N_j}{\partial y}\right)
$$

$$
k_2ij = \sum\sum W\left(D_2\frac{\partial N_i}{\partial x}\frac{\partial N_j}{\partial x} + D_5\frac{\partial N_i}{\partial y}\frac{\partial N_j}{\partial y}\right)
$$

$$
\tag{3.29}
$$

$$
k_3ij = \sum\sum W\left(D_2\frac{\partial N_i}{\partial y}\frac{\partial N_j}{\partial x} + D_5\frac{\partial N_i}{\partial x}\frac{\partial N_j}{\partial y}\right)
$$

$$
k_4ij = \sum\sum W\left(D_4\frac{\partial N_i}{\partial y}\frac{\partial N_j}{\partial y} + D_5\frac{\partial N_i}{\partial x}\frac{\partial N_j}{\partial x}\right)
$$

By using these explicit expressions instead of carrying out the full
matrix multiplication for the triple product, it is possible to reduce
the element stiffness formulation time by a factor of four.

A close inspection of Eqn.(3.29) shows that all the terms are of the
form (constant) × (material property coefficient) × (product of shape
function derivatives). Therefore, providing that the material properties
are constant throughout the element, further economies can be introduced
into the computation by storing separately in an array $[S]_{ij}$ the sum of
the products of the shape function derivatives multiplied by W, such
that

$$
[S]_{ij} =
\begin{bmatrix}
\displaystyle\sum\sum W \frac{\partial N_i}{\partial x}\frac{\partial N_j}{\partial x} & \displaystyle\sum\sum W \frac{\partial N_i}{\partial x}\frac{\partial N_j}{\partial y} \\[2em]
\displaystyle\sum\sum W \frac{\partial N_i}{\partial y}\frac{\partial N_j}{\partial y} & \displaystyle\sum\sum W \frac{\partial N_i}{\partial y}\frac{\partial N_j}{\partial x}
\end{bmatrix}
\tag{3.30}
$$

and then to work out Eqn.(3.29) by incorporating the relevant material
constants.

This method, which was first reported by Gupta [5], will cut the
computing time for element stiffness formulation by a factor of nine
over the full matrix multiplication method, and is the method implemented in the illustrative program to be shown later. For problems
involving variable material properties within the elements, however,
the Gupta algorithm cannot be used.

In the illustrative program the element stiffness formulation is
controlled by subroutine STIFN which sets up the geometry of each
element in turn and then carries out the numerical integration.

Inside the innermost loop of the numerical integration there is a call
to subroutine JACOB which, for the Gauss point under consideration at
that time, calculates the values of the shape function, the shape
function derivatives and $\det[J]$.

Immediately after the call to JACOB there is a call to MULT1 which
assembles, in the element stiffness, the product of the shape functions
as per Eqn.(3.30). After the Gauss loops have been completed MULT2 is
called which carries out the rearrangement to produce the element
stiffness as per Eqn.(3.29).

Subroutine STIFN (maxmat, maxnel, rs, s, mat,
nnodz, ldef, ldest, cord, nrule
vectlc, wtfun

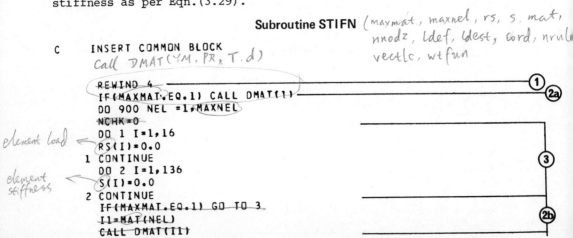

```
C      INSERT COMMON BLOCK
       Call DMAT(YM, PR, T, d)

       REWIND 4                                          ①
       IF(MAXMAT.EQ.1) CALL DMAT(1)                     ②a
       DO 900 NEL =1,MAXNEL
       NCHK=0
       DO 1 I=1,16          ← element load
       RS(I)=0.0
    1  CONTINUE                                          ③
       DO 2 I=1,136         ← element stiffness
       S(I)=0.0
    2  CONTINUE
       IF(MAXMAT.EQ.1) GO TO 3
       I1=MAT(NEL)
       CALL DMAT(I1)                                     ②b
```

number of nodes per element

```
      DO 10 J=1,NNODZ
      NIC=IABS(LDEF(NEL,J))
      NELDES(J)=LDEST(NIC)
      NELDEF(J)=NIC
      X(J)=CORD(NIC,1)
      Y(J)=CORD(NIC,2)
   10 CONTINUE
      CALL INTCRD
      IF(MAXRCT.NE.0) CALL PLOAD
      DO 100 JA=1,NRULE
      XL=VECTLC(JA)
      WX=WTFUN(JA)
      DO 90 JB=1,NRULE
      YL=VECTLC(JB)
      WY=WTFUN(JB)
      CALL JACOB(0)
      CALL MULT1
      IF(GRAVX.NE.0.0.OR.GRAVY.NE.0.0) CALL GRAV
      IF(WIRL.NE.0) CALL ROTAT
   80 CONTINUE
   90 CONTINUE
  100 CONTINUE
      CALL MULT2
      IF(MAXPRS.NE.0) CALL PRES1
      IF(MAXINS.NE.0) CALL TRANSF
      WRITE(4) S,RS,(LDEF(NEL,J),J=1,NNODZ),NELDES
  900 CONTINUE
      IF(NSTOP.EQ.0) RETURN
      WRITE(6,1000)
      STOP
 1000 FORMAT( 50H1ILLCONDITIONING OR GEOMETRY OR DEFINITION ERRORS.,/,
     1 22H EXECUTION TERMINATED.)
      END
```

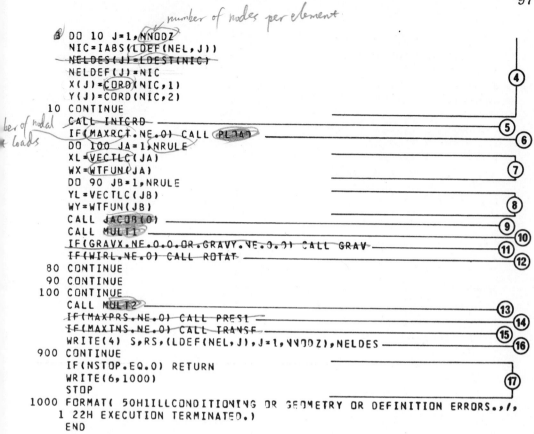

number of nodal point loads

(1) Tape 4 is reset to receive the element stiffness and load vectors for reading back in subroutine FRONT.
(2) Subroutine DMAT sets up the material property matrix for the element. If all elements are made up of the same material DMAT is called once only (2a) or once for each element (2b).
(3) The element load vector RS and the element stiffness S are initialized.
(4) The element destination vector NELDES, the element definition vector NELDEF and the nodal coordinate vectors X and Y are set up.
(5) Subroutine INTCRD checks for any mid-side nodal coordinates that have not been defined. These coordinates are then defined by a linear interpretation between the appropriate corner nodes.
(6) If there are any nodal point loads subroutine PLOAD is called to check for any such loads in the element under consideration.
(7) XL is the ξ coordinate of each Gauss point in turn and WX the associated weighting function.
(8) YL is the η coordinate of each Gauss point in turn and WY the associated weighting function.
(9) Subroutine JACOB is called to calculate the shape function and the shape function derivatives.
(10) Subroutine MULT1 assembles into the element stiffness vector the products of the shape function derivatives and the weighting function as per equations (3.30).
(11) Subroutine GRAV is called if there are any gravity loads.
(12) Subroutine ROTAT is called if there are any rotational loads.

(13) Subroutine MULT2 is called to calculate the actual element stiffness from the product of shape function derivatives etc. as set up by MULT1.

(14) Subroutine PRES1 is called if there are any pressure loads.

(15) Subroutine TRANSF is called if there are any nodes which have to have their displacements defined relative to a local set of axes and not the global axis.

(16) The element stiffness, load vector, definition vector and destination vector are written onto tape 4.

(17) If badly distorted elements have been encountered, indicated by DETJ<0, then execution stops.

Subroutine MULT1 (WX, WY, DETJ, nnodz, DX, DY, S∅, F)

```
C      INSERT COMMON BLOCK
       dimension DX( 8 ) , DY( 8 ) , S( 4 )
       do 10 i=1, 4
    10 S(i)=0.0    nue
       N=0
       WAIT=WX*WY*DETJ*T                          ──① 
       DO 20 I=1, NNODZ
       DXI=DX(I)*WAIT                              ──②
       DYI=DY(I)*WAIT
       DO 15 J=I, NNODZ
       DXJ=DX(J)  ①  ?
       DYJ=DY(J)                                   ──③
       S(N+1)=S(N+1)+DXI*DXJ
       S(N+2)=S(N+2)+DXI*DYJ                       ──④
       S(N+3)=S(N+3)+DYI*DYJ
       N=N+3
       IF(I.EQ.J) GO TO 15
       S(N+1)=S(N+1)+DYI*DXJ                       ──⑤
       N=N+1
    15 CONTINUE
    20 CONTINUE
       RETURN
       END
```

(1) WAIT = $W_i * W_j * \det[J]$

(2) DX(I) = $\dfrac{\partial N_i}{\partial x}$

   DY(I) = $\dfrac{\partial N_i}{\partial y}$

(3) DX(J) = $\dfrac{\partial N_j}{\partial x}$

   DY(J) = $\dfrac{\partial N_j}{\partial y}$

(4) Setting up k_1, k_2 and k_3 as per Eqn.(3.30)

(5) If I $\neq$ J then k_4 is set up.

Subroutine MULT2 *(nnodz, s, d)*

```
C     INSERT COMMON BLOCK
      dimension s(4), d(5)

      N=0
      DO 40 I=1,NNODZ
      DO 30 J=I,NNODZ
      S1=S(N+1)
      S2=S(N+2)
      S3=S(N+3)
      IF(I.EQ.J) GO TO 20
      S4=S(N+4)
      S(N+1)=S1*D(1)+S3*D(5)
      S(N+2)=S2*D(2)+S4*D(5)
      S(N+3)=S4*D(2)+S2*D(5)
      S(N+4)=S3*D(4)+S1*D(5)
      N=N+4
      GO TO 30
   20 S(N+1)=S1*D(1)+S3*D(5)
      S(N+2)=S2*(D(2)+D(5))
      S(N+3)=S3*D(4)+S1*D(5)
      N=N+3
   30 CONTINUE
   40 CONTINUE
      RETURN
      END
```

(1) The products of the shape function derivatives and weighting functions are retrieved from the element stiffness.
(2) If I ≠ J then the 2 × 2 submatrix element stiffness is calculated by multiplying the products of the shape functions etc. by coefficients of the material property matrix.
(3) If I = J then the upper triangle of the 2 × 2 submatrix element stiffness is calculated.

Subroutine DMAT(I1) *(YM, PR, T, d).*
dimension YM(maxmat), PR(maxmat), T(maxmat), d(5)

```
C     INSERT COMMON BLOCK
      do 10 i=1, 5
      d(i)=0.0
   10 continue
      YMOD=YM(I1)
      PRAT=PR(I1)
THICK=T(I1)  WEIGHT=WT(I1)
      D(1)=YMOD/(1.0-PRAT*PRAT)
      D(2)=D(1)*PRAT
      D(3)=D(2)
      D(4)=D(1)
      D(5)=YMOD/(2.0*(1.0+PRAT))
      RETURN
      END
```

(1) Young's modulus.
(2) Poisson's ratio.
(3) Mass/unit area of material forming the element.
(4) The coefficients of the material property matrix are calculated.

Subroutine JACOB (I1) *(CW, nnodz, XX, YY, X, Y, DX, D*
DETJ, AW)

dimension CW(4), AW(4), XX(8), YY(8), X(8), Y(8)

```
C     INSERT COMMON BLOCK    DX(8), DY(8)
                do 6, i=1,8
                   DX(i)=0.0  DY(i)=0.0, SHP(i)=0.0.
          6    Continue
              DO 5 I=1,4
              CW(I)=0.0 , AW(I)=0.0
        5 CONTINUE
              DO 100 I=1,NNODZ
              XXI=XX(I) ξi
              YYI=YY(I) ηi
              GO TO (10,20,10,30,10,20,10,30),I
       10 A1=XL*XXI
          A2=YL*YYI
          DX(I)=0.25*XXI*(1.0+A2)*(2.0*A1+A2)
          DY(I)=0.25*YYI*(1.0+A1)*(2.0*A2+A1)
          SHP(I)=0.25*(1.0+A1)*(1.0+A2)*(A1+A2-1.0)  → Pg 88.
          GO TO 90
       20 DX(I)=0.5*XXI*(1.0-YL*YL)
          DY(I)=-YL*(1.0+XL*XXI)
          SHP(I)=0.5*(1.0+XL*XXI)*(1.0-YL*YL)
          GO TO 90
       30 DX(I)=-XL*(1.0+YL*YYI)
          DY(I)=0.5*YYI*(1.0-XL*XL)
          SHP(I)=0.5*(1.0+YL*YYI)*(1.0-XL*XL)
       90 CW(1)=CW(1)+DX(I)*X(I)
          CW(2)=CW(2)+DX(I)*Y(I)
          CW(3)=CW(3)+DY(I)*X(I)
          CW(4)=CW(4)+DY(I)*Y(I)
      100 CONTINUE
          DETJ=CW(1)*CW(4)-CW(2)*CW(3)
          IF(DETJ.GT.0.0) GO TO 110
          WRITE(6,1000) NEL,DETJ
          NSTOP=1
      110 AW(1)=CW(4)
          AW(2)=-CW(2)
          AW(3)=-CW(3)
          AW(4)=CW(1)
          IF(I1.EQ.1) RETURN
          DO 120 I=1,4
          AW(I)=AW(I)/DETJ
      120 CONTINUE
          DO 200 I=1,NNODZ
          DXI=DX(I)
          DYI=DY(I)
          DX(I)=AW(1)*DXI+AW(2)*DYI
          DY(I)=AW(3)*DXI+AW(4)*DYI
      200 CONTINUE
          RETURN
     1000 FORMAT(//,5X,28HNEGATIVE OR ZERO DETERMINENT,/,5X,8HELEMENT ,I5,
        1        12H DETERMINENT,2X,E10.3)
          END
```

(1) The vector CW that will eventually contain the Jacobian matrix is initialized.
(2) (XXI, YYI) the local coordinates of each of the nodes in turn.
(3) The equation to be used to calculate the shape function and its derivatives with respect to ξ,η depends upon which node is being considered.
(4) Corner nodes

$$DX(I) = \frac{\partial N_i}{\partial \xi} \qquad DY(I) = \frac{\partial N_i}{\partial \eta}$$

(5) Midside nodes η = 0.
(6) Midside nodes ξ = 0.
(7) The Jacobian matrix is assembled as

$$[J] = \begin{bmatrix} CW(1) & CW(2) \\ CW(3) & CW(4) \end{bmatrix}$$

(8) $\det[J]$.
(9)
$$[J]^{-1} = \begin{bmatrix} AW(1) & AW(2) \\ AW(3) & AW(4) \end{bmatrix}$$

(10) Derivatives of shape function with respect to x,y:

$$DX(I) = \frac{\partial N_i}{\partial x}$$

$$DY(I) = \frac{\partial N_i}{\partial y}$$

3.7 Stress matrix

The calculation of the stresses in an element is a simple matter once the nodal displacements have been computed. Eqn.(1.38) is used with the displacement vector containing the nodal displacements for the element considered.

The explicit form of Eqn.(1.38) for the quadratic element is

$$\begin{Bmatrix} \sigma_x \\ \sigma_y \\ \tau_{xy} \end{Bmatrix} = \begin{bmatrix} D_1 & D_2 & 0 \\ D_2 & D_4 & 0 \\ 0 & 0 & D_5 \end{bmatrix} \begin{bmatrix} \frac{\partial N_1}{\partial x} & 0 & \frac{\partial N_2}{\partial x} & 0 & \cdots & 0 \\ 0 & \frac{\partial N_1}{\partial y} & 0 & \frac{\partial N_2}{\partial y} & \cdots & \frac{\partial N_8}{\partial y} \\ \frac{\partial N_1}{\partial y} & \frac{\partial N_1}{\partial x} & \frac{\partial N_2}{\partial y} & \frac{\partial N_2}{\partial x} & \cdots & \frac{\partial N_8}{\partial x} \end{bmatrix} \begin{Bmatrix} u_1 \\ v_1 \\ u_2 \\ \vdots \\ v_8 \end{Bmatrix} \tag{3.31a}$$

or in longhand form,

$$\sigma_x = \left[D_1 \sum_{i=1}^{8} \frac{\partial N_i}{\partial x} u_i + D_2 \sum_{i=1}^{8} \frac{\partial N_i}{\partial y} v_i \right]$$

$$\sigma_y = \left[D_2 \sum_{i=1}^{8} \frac{\partial N_i}{\partial x} u_i + D_4 \sum_{i=1}^{8} \frac{\partial N_i}{\partial y} v_i \right] \tag{3.31b}$$

$$\tau_{yx} = D_5 \left[\sum_{i=1}^{8} \left(\frac{\partial N_i}{\partial y} u_i + \frac{\partial N_i}{\partial x} v_i \right) \right]$$

For computational efficiency it is preferable to use Eqn.(3.31b). In the illustrative program subroutine STRESS controls the stress calculation. It sets up the geometry of each element in turn and loops through each of the element nodes. For each node it sets up the ξ,η coordinates and calls JACOB to determine the shape function derivatives. The calculation of stresses as per Eqn.(3.31b) is carried out by subroutine MULT3.

In a number of programs the stresses are determined at the nodes, since the nodal positions are readily located and it is convenient to output the displacements and stresses at the same points. It has been found that nodal stresses from a quadratic element are usually incorrect, however if the stresses of all elements meeting at a node are averaged much closer answers are obtained.

A better alternative is to calculate the stresses at the Gauss points, in which it will be found that because of the superior accuracy no averaging is necessary, or indeed, is possible. The stresses can be presented as stress contours; or a least square smoothing technique is used to yield a surface function from which accurate nodal stresses are obtained.

One explanation for the Gauss point stresses being more accurate than the nodal stresses is that the element stiffness is calculated by sampling at the Gauss points and it is therefore reasonable to expect the most accurate stresses and strains at the same points.

Subroutine STRESS

```
C     INSERT COMMON BLOCK


      WRITE(6,1000)                                         ──┐
      DO 10 I=1,MAXNOD                                        │
      WRITE(6,1001) I,(DISPL(I,J),J=1,NVABZ)                 ①
   10 CONTINUE                                              ──┘
      IF(NPUT.EQ.0) GO TO 20 ──────────────────────────────②
      N1=MAXNOD*4   (stress + 1 counter)                   ──┐
      IF(N1.LE.MAXSS) GO TO 11                               │
      NPUT=0                                                 ③
      WRITE(6,2000)                                          │
      GO TO 20                                              ──┘
   11 DO 12 I=1,N1                                          ──┐
      SS(I)=0.0                                              ④
   12 CONTINUE                                              ──┘
   20 DO 100 NEL=1,MAXNEL                                  ──┐
      IF(MAXMAT.EQ.1) GO TO 25                               │
      N2=MAT(NEL)                                            ⑤
      CALL DMAT(N2)                                         ──┘
   25 WRITE(6,1002) NEL                                    ──┐
      DO 30 NOD=1,NNODZ                                      │
      N3=IABS(LDEF(NEL,NOD))                                 │
      NELDEF(NOD)=N3                                         │
      X(NOD)=CORD(N3,1)                                      │
      Y(NOD)=CORD(N3,2)                                      ⑥
      U(NOD)=DISPL(N3,1)                                     │
      V(NOD)=DISPL(N3,2)                                    ──┘
   30 CONTINUE                                             ──┐
      DO 50 NOD=1,NNODZ                                      │
      XL=XX(NOD)                                             ⑦
      YL=YY(NOD)                                            ──┘
```

```
      CALL JACOB(0)
      CALL MULT3
      NIC=NELDEF(NOD)
      WRITE(6,1003) NIC,SIGMA
      IF(NPUT.EQ.0) GO TO 50
      N4=(NIC-1)*4
      DO 40 I=1,3
      SS(N4+I)=SS(N4+I)+SIGMA(I)
   40 CONTINUE
      SS(N4+4)=SS(N4+4)+1.0
   50 CONTINUE
  100 CONTINUE
      IF(NPUT.EQ.0) RETURN
      N5=-4
      WRITE(6,1004)
      DO 150 NOD=1,MAXNOD
      N5=N5+4
      DIV=SS(N5+4)
      IF(DIV.EQ.0.0) GO TO 150
      DO 140 I=1,3
      SIGMA(I)=SS(N5+I)/DIV
  140 CONTINUE
      WRITE(6,1003) NOD,SIGMA
  150 CONTINUE
      RETURN

 1000 FORMAT(23H1NODAL DISPLACEMENTS,/,
     1       7X,4HNODE,3X,6HX-COMP,9X,6HY-COMP)
 1001 FORMAT(6X,I5,3(5X,F10.7))
 1002 FORMAT(//////,17H STRESSES ELEMENT,I4,//,
     1       7X,4HNODE,6X,9HSIGMA X-X,3X,9HSIGMA Y-Y,5X,7HTAU X-Y)
 1003 FORMAT(/,1X,I10,3X,6(2X,F10.1))
 1004 FORMAT(23H1AVERAGE NODAL STRESSES,//,
     1       7X,4HNODE,6X,9HSIGMA X-X,3X,9HSIGMA Y-Y,5X,7HTAU X-Y)
 2000 FORMAT( 65H1INSUFFICIENT SPACE IN SS VECTOR TO ALLOW NODAL STRESS
     1AVERAGING.,/, 21H AVERAGING CANCELLED.)
      END
```

(1) The nodal displacements as calculated by the solution subroutines are printed out.

(2) If nodal stress averaging is requested, NPUT=1, then the stresses and the counter, used to indicate the number of elements connected to each node, are stored in the structural stiffness vector SS.

(3) If there is insufficient room in SS then nodal stress averaging is cancelled.

(4) The SS vector is initialized if nodal stress averaging is requested.

(5) The material property matrix for the element is set up.

(6) The element definition vector NELDEF, the nodal global coordinate vectors, X,Y and the nodal displacement vectors U,V are set up.

(7) The local coordinates of each of the nodes in turn XL = ξ, YL = η.

(8) Subroutine JACOB calculates the shape function derivatives for the node defined by (XL,YL).

(9) MULT3 determines the nodal stress as per Eqn.(3.31b).

(10) The nodal stresses are printed out.

(11) If nodal stress averaging requested the nodal stresses are stored in the SS vector.

(12) The counter is incremented.

(13) The average nodal stresses are calculated and printed out.

Subroutine MULT3

```
C     INSERT COMMON BLOCK

      DO 1 I=1,4
      CW(I)=0.0
    1 CONTINUE
      DO 10 I=1,NNODZ
      CW(1)=CW(1)+DX(I)*U(I)
      CW(2)=CW(2)+DX(I)*V(I)
      CW(3)=CW(3)+DY(I)*U(I)
      CW(4)=CW(4)+DY(I)*V(I)
   10 CONTINUE
      SIGMA(1)=CW(1)*D(1)+CW(4)*D(2)
      SIGMA(2)=CW(1)*D(3)+CW(4)*D(4)
      SIGMA(3)=D(5)*(CW(2)+CW(3))
      RETURN
      END
```

(1) The CW vector is initialized.
(2) The products of the shape function derivatives and the nodal
    displacements are stored in CW.
(3) The nodal stresses are calculated as per Eqn.(3.31b).

3.8 Some consistent load matrices

The types of loading that are applied to a two-dimensional element can
be divided into two groups: edge loads such as pressure and nodal points
loads, the body forces such as those due to gravity and centrifugal
loads.

Unlike the triangular element, in which all loads can be assigned to
nodes intuitively or by statics, for a quadratic isoparametric the nodal
loads due to distributed loads must be computed in accordance with Eqn.
(1.47). The equivalent nodal forces are added, element by element, into
a load vector to be used in the solution procedure. The treatment of
the different type of loads is discussed in detail in the following
subsections.

3.8.1 NODAL POINT LOADS

It is usually assumed that point loads will always be applied at nodes
and not at an arbitrary point on an element boundary. This requirement
is easily accommodated by putting in nodes at the points where the
concentrated loads are acting.

For the front solver to be described in Chapter 4, the assembly is
independent of the numbering of nodes, and at any one time, only the
stiffness and the load coefficients connected with the front are stored.
It is therefore more convenient to treat a nodal point load as a concentrated load acting on only one of the elements which are connected
to the node in question, and then the nodal load vectors for all types
of loading will be calculated on an element basis and assembled by the
subroutine ASMBLE. Subroutine PLOAD determines the nodal point loads.

3.8.2 EDGE PRESSURE

It is usual to disregard the actual pressure variation along an element edge but simply to assume a parabolic distribution defined by the pressure values at each of the three nodes along that edge, and all intermediate pressure values can be calculated using the shape functions. For convenience in coding, all the nodes of the element are used in the computation so that there is no need to sort out the appropriate shape functions for the three nodes with given pressure values, which can include an input value of zero.   Thus

$$P = \sum_{k=1}^{8} N_k P_k \qquad\qquad (3.32)$$

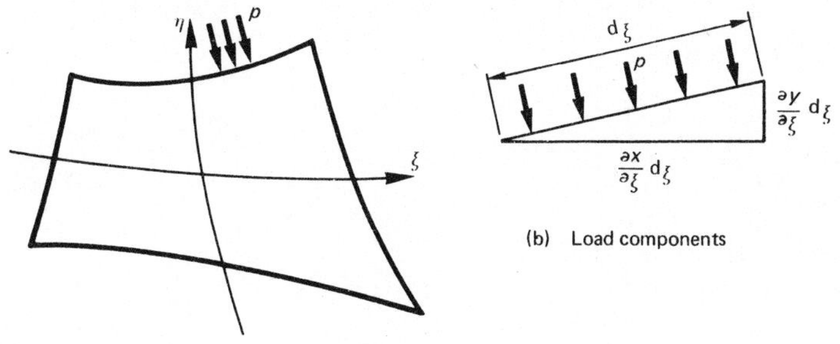

(b)   Load components

(a)   Element with edge loading

Fig.3.6 Edge pressure for an isoparametric element

For example, if we consider a pressure distributed in some manner along the $\eta = +1$ edge of an element (Fig.3.6), the components of the force $p\,d\xi$ acting upon an elemental length $d\xi$ are

$$\begin{Bmatrix} P_x \\ P_y \end{Bmatrix} = P \begin{Bmatrix} \dfrac{\partial y}{\partial \xi} \\ -\dfrac{\partial x}{\partial \xi} \end{Bmatrix} d\xi \qquad\qquad (3.33)$$

p is specified in force per unit length since $d\xi$ (*see* Eqn.(3.18) is in fact defined with respect to x,y and not to the dimensionless ξ,η coordinates.

The consistent load for p is given by a modified form of Eqn.(1.47), in which the volume integral has been reduced to a line integral.

$$\{P\} = \int_{-1}^{+1} P[N]^T \begin{Bmatrix} \dfrac{\partial y}{\partial \xi} \\ -\dfrac{\partial x}{\partial \xi} \end{Bmatrix} d\xi \qquad\qquad (3.34a)$$

where

$$[N] = \begin{bmatrix} N_1 & 0 & N_2 & 0 & \cdots & 0 \\ 0 & N_1 & 0 & N_2 & \cdots & N_8 \end{bmatrix} \qquad\qquad (3.34b)$$

The above equation is usually integrated numerically and hence needs to be rewritten as

$$\{P\} = \sum_{i=1}^{n} W_i [N]^T p_i \left\{ \begin{array}{c} \dfrac{\partial y}{\partial \xi} \\[2ex] -\dfrac{\partial x}{\partial \xi} \end{array} \right\}_i \qquad (3.34c)$$

where p_i is the pressure at the Gauss point i along the $\eta = +1$ edge and is computed by Eqn.(3.32).

If the same pressure acts on the edge $\eta = -1$ the sign of the load vector $\{P\}$ would be reversed since a positive pressure is assumed to act towards the centre of an element. A more general equation applicable to the two edges $\eta = \pm 1$ is

$$\{P\} = \sum_{i=1}^{n} W_i W_\eta [N]^T \left(\sum_{k=1}^{8} N_k p_k \right) \left\{ \begin{array}{c} \dfrac{\partial y}{\partial \xi} \\[2ex] -\dfrac{\partial x}{\partial \xi} \end{array} \right\} \qquad (3.35a)$$

where W_η takes up the value of the η coordinate for the loaded edge. A similar expression for pressure loads on the $\xi = \pm 1$ edges is

$$\{P\} = \sum_{i=1}^{n} W_i W_\xi [N]^T \left(\sum_{k=1}^{8} N_k p_k \right) \left\{ \begin{array}{c} -\dfrac{\partial y}{\partial \eta} \\[2ex] \dfrac{\partial x}{\partial \eta} \end{array} \right\} \qquad (3.35b)$$

Since the location of the Gauss points along the edge are not the same as those used to calculate the element stiffness it is usual to call the pressure subroutines from STIFN (Section 3.6) after the computation of element stiffness matrix. PRES1 determines whether an element has a pressure loading and also identifies the edge on which the pressure acts. An edge is only considered to be pressure loaded if the pressure values (zero or non-zero) for all three nodes are defined. PRES1 then calls PRES2 to determine the value of W_η or W_ξ, which in turn will call JACOB to evaluate $[N]$ and $[J]$ for each of the Gauss points, and then PRES3 to determine p and to compute the load vector from Eqn.(3.35).

3.8.3 CENTRIFUGAL FORCES

For ease of programming it is assumed that the centre of rotation coincides with the origin of the x, y axes.

The radially outward body force p_r on an elemental area dA is

$$p_r = r\Omega^2 m \qquad (3.36)$$

in which

Ω is the angular velocity in radians per second;
m is the mass per unit area of the material;
r is the radial distance from the origin to the centroid of the elemental area.

p_r can be resolved into components parallel to x and y axes such that

$$\begin{Bmatrix} P_x \\ P_y \end{Bmatrix} = \Omega^2 m \begin{Bmatrix} x \\ y \end{Bmatrix} \tag{3.37}$$

where x and y correspond to the coordinates of the centroid of the elemental area.

By virtue of Eqn.(1.47), the equivalent nodal forces are

$$\{P\} = \int \Omega^2 m [N]^T \begin{Bmatrix} x \\ y \end{Bmatrix} dx\,dy \tag{3.38}$$

Once again the integration is carried out numerically and Eqn.(3.38) takes up the form of

$$\{P\} = \int_{-1}^{+1}\int_{-1}^{+1} \Omega^2 m [N]^T \begin{Bmatrix} x \\ y \end{Bmatrix} \det[J]\,d\xi\,d\eta$$

$$= \sum_{i=1}^{n}\sum_{j=1}^{n} W_i W_j \Omega^2 m [N]^T \begin{Bmatrix} x_{ij} \\ y_{ij} \end{Bmatrix} \det[J] \tag{3.39}$$

x_{ij} and y_{ij} are the x and y coordinates of each of the Gauss points respectively, and are easily computed with the help of Eqns.(3.6b) and (3.6c):

$$\begin{Bmatrix} x_{ij} \\ y_{ij} \end{Bmatrix} = \sum_{k=1}^{8} \begin{Bmatrix} N_k(\xi_i,\eta_j)x_k \\ N_k(\xi_i,\eta_j)y_k \end{Bmatrix} \tag{3.40}$$

It is natural to use the same Gauss points for computing the stiffness matrix and the equivalent load vector, and subroutine ROTAT is called inside the inner loop of STIFN for each Gauss point in turn to work out Eqn.(3.39).

3.8.4 BODY FORCES

The usual types of body forces are those due to gravity (g_y) and those due to earthquake loading (g_x). The formulation is identical to that for centrifugal forces and we have

$$\{P\} = \int_{-1}^{+1}\int_{-1}^{+1} m[N]^T \begin{Bmatrix} g_x \\ g_y \end{Bmatrix} \det[J] d\xi d\eta$$

$$= \sum_{i=1}^{n}\sum_{j=1}^{n} W_i W_j m[N]^T \begin{Bmatrix} g_x \\ g_y \end{Bmatrix} \det[J] \qquad (3.41)$$

where m is the mass per unit area and g_x and g_y the accelerations in the x and y directions respectively. The subroutine GRAV for computing Eqn. (3.41) is also called inside the inner loop of STIFN.

Subroutine PLOAD

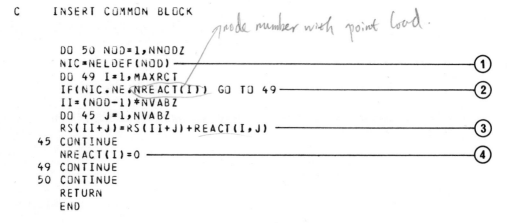

```
C     INSERT COMMON BLOCK
                                       node number with point load.
      DO 50 NOD=1,NNODZ
      NIC=NELDEF(NOD) ─────────────────────────────────── ①
      DO 49 I=1,MAXRCT
      IF(NIC.NE.NREACT(I)) GO TO 49 ─────────────────────── ②
      II=(NOD-1)*NVABZ
      DO 45 J=1,NVABZ
      RS(II+J)=RS(II+J)+REACT(I,J) ──────────────────────── ③
   45 CONTINUE
      NREACT(I)=0 ───────────────────────────────────────── ④
   49 CONTINUE
   50 CONTINUE
      RETURN
      END
```

(1) Each node in the element is taken in turn and
(2) compared with each node that has a point load applied.
(3) If a node has an applied point load then that load is assembled into the element load vector.
(4) To ensure that the point loads at the node are not assembled into another element load vector, of an element containing that node, the node number in the loaded node list NREACT is reset to zero.

Subroutine PRES1

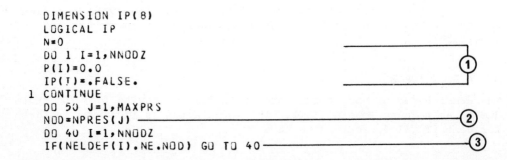

```
C     INSERT COMMON BLOCK

      DIMENSION IP(8)
      LOGICAL IP
      N=0
      DO 1 I=1,NNODZ
      P(I)=0.0                                               ①
      IP(I)=.FALSE.
    1 CONTINUE
      DO 50 J=1,MAXPRS
      NOD=NPRES(J) ──────────────────────────────────────── ②
      DO 40 I=1,NNODZ
      IF(NELDEF(I).NE.NOD) GO TO 40 ─────────────────────── ③
```

```
      P(I)=PRES(J)
      N=N+1
      IP(I)=.TRUE.
      GO TO 50
   40 CONTINUE
   50 CONTINUE
      IF(N.LT.3) RETURN
      IF(IP(1).AND.IP(2).AND.IP(3)) CALL PRES2(1)
      IF(IP(5).AND.IP(6).AND.IP(7)) CALL PRES2(2)
      IF(IP(3).AND.IP(4).AND.IP(5)) CALL PRES2(3)
      IF(IP(7).AND.IP(8).AND.IP(1)) CALL PRES2(4)
      RETURN
      END
```

(1) IP and P are initialized. IP indicates if a node has a pressure
    applied. P indicates the pressure at the node.
(2) Each node in the pressure loaded node list is taken in turn and
(3) compared with each of the nodes forming the element.
(4) If an element node is pressure loaded then the pressure is stored
    in P and the indicator IP is set to true.
(5) If there are less than three nodes loaded no edge has a pressure
    load.
(6) Each of the element edges is considered in turn and PRES2 called
    if an edge is pressure loaded.

Subroutine PRES2(I1)

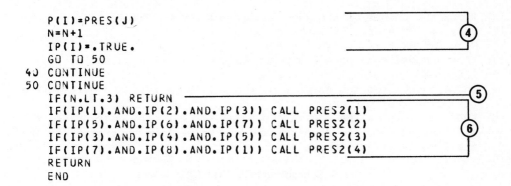

```
C     INSERT COMMON BLOCK


      GO TO(50,52,100,102),I1
   50 WX=1.0
      XL=1.0
      GO TO 53
   52 WX=-1.0
      XL=-1.0
   53 DO 60 JA=1,NRULE
      YL=VECTLC(JA)
      WY=WTFUN(JA)
      CALL JACOB(1)
      CALL PRES3(1,3)
   60 CONTINUE
      RETURN
  100 WY=1.0
      YL=1.0
      GO TO 103
  102 WY=-1.0
      YL=-1.0
  103 DO 120 JA=1,NRULE
      XL=VECTLC(JA)
      WX=WTFUN(JA)
      CALL JACOB(1)
      CALL PRES3(2,4)
  120 CONTINUE
      RETURN
      END
```

(1) Il is the number of the edge loaded defined by PRES1.
(2) Edge 1: WX = W_ξ,   XL = ξ = 1.
(3) Edge 2: WX = W_ξ,   XL = ξ = - 1.
(4) YL = η coordinate of Gauss point in turn and WY = weighting function.
(5) JACOB calculates the shape function and shape function derivatives for point (XL,YL).
(6) PRES3 calculates the equivalent nodal force to the edge pressure.
(7)-(11) For edges 3 and 4, η = 1 and - 1 respectively, procedures are the same as (1)-(6) above.

Subroutine PRES3(I1,I2)

```
C      INSERT COMMON BLOCK


       APRES=0.0
       DO 10 I=1,NNODZ
       APRES=APRES+P(I)*SHP(I) ──────────────────────────(1)
    10 CONTINUE
       WAIT=WX*WY*APRES ─────────────────────────────────(2)
       M=0
       DO 20 I=1,NNODZ
       SHPI=SHP(I)*WAIT ──────────────────────────(
       RS(M+1)=RS(M+1)-SHPI*AW(I1)                       (4)
       RS(M+2)=RS(M+2)-SHPI*AW(I2)  ──────────────────
       M=M+NVABZ
    20 CONTINUE
       RETURN
       END
```

(1) APRES $= \sum N_i P_i$

(2) WAIT $= W_i *$ (W_ξ) $* \left(\sum N_i P_i\right) * \det[J]$
                or
                (W_η)

(3) SHPI $= N_i * W_i *$ (W_ξ) $* \left(\sum N_i P_i\right) * \det[J]$
                    or
                    (W_η)

(4) The equivalent nodal loads are assembled as per Eqns.(3.35a) or (3.35b) depending upon which edge is loaded.

Subroutine GRAV

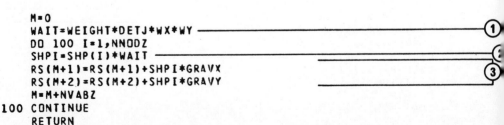

```
C      INSERT COMMON BLOCK


       M=0
       WAIT=WEIGHT*DETJ*WX*WY ───────────────────────────(1)
       DO 100 I=1,NNODZ
       SHPI=SHP(I)*WAIT ──────────────────────────(
       RS(M+1)=RS(M+1)+SHPI*GRAVX                        (3)
       RS(M+2)=RS(M+2)+SHPI*GRAVY  ──────────────────
       M=M+NVABZ
   100 CONTINUE
       RETURN
       END
```

(1) WAIT = $M * \det[J] * W_i * W_j$

(2) SHPI = $N_i * M * \det[J] * W_i * W_j$

(3) The equivalent nodal loads are assembled as per Eqn.(3.41).

Subroutine ROTAT

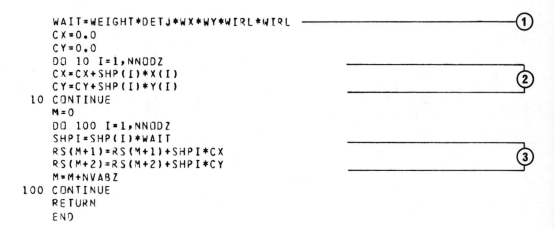

```
C     INSERT COMMON BLOCK

      WAIT=WEIGHT*DETJ*WX*WY*WIRL*WIRL                              ①
      CX=0.0
      CY=0.0
      DO 10 I=1,NNODZ
      CX=CX+SHP(I)*X(I)                                            ②
      CY=CY+SHP(I)*Y(I)
   10 CONTINUE
      M=0
      DO 100 I=1,NNODZ
      SHPI=SHP(I)*WAIT
      RS(M+1)=RS(M+1)+SHPI*CX                                      ③
      RS(M+2)=RS(M+2)+SHPI*CY
      M=M+NVABZ
  100 CONTINUE
      RETURN
      END
```

(1) WAIT = $W_i W_j \Omega^2 m \det[J]$

(2) $\left.\begin{array}{l} CX \\ CX \end{array}\right\}$ Global coordinates of the Gauss point.

(3) Equivalent nodal loads calculated as per Eqn.(3.39) and assembled into the element load vector.

3.8.5 EXAMPLES OF LOADING

For a rectangular isoparametric element with edges parallel to the x,y axes it is possible to integrate the equations for the equivalent nodal loads explicitly. For example, consider the element of Fig.3.7 with a uniform pressure load along the top edge.

Using Eqns.(3.9) and (3.34a) and noting that for edge $\eta = 1$, $N_1 = N_2 = N_3 = N_4 = N_8 = 0$ and for the rectangular element, $\partial y/\partial \xi = 0$ and $\partial x/\partial \xi = a$, we find that

$$\{P\} = \begin{Bmatrix} P_{y5} \\ P_{y6} \\ P_{y7} \end{Bmatrix} = \int_{-1}^{1} p \begin{Bmatrix} N_5 \\ N_6 \\ N_7 \end{Bmatrix} (-a)\,\mathrm{d}\xi$$

Writing $p(2a) = W$, the total pressure load, then

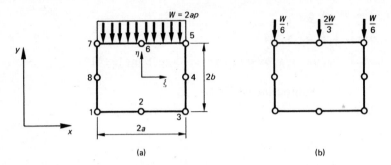

Fig.3.7 Edge loading and equivalent nodal forces

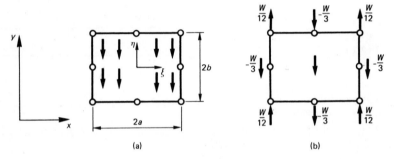

Fig.3.8 Gravity loading and equivalent nodal forces

$$P_{y5} = \frac{-W}{2} \int_{-1}^{1} \frac{1}{4}(1 + \xi\xi_i)(1 + \eta\eta_i)(\xi\xi_i + \eta\eta_i - 1)\,d\xi \qquad (3.42a)$$

$$P_{y6} = \frac{-W}{2} \int_{-1}^{1} \frac{1}{2}(1 + \eta\eta_i)(1 - \xi^2)\,d\xi \qquad (3.42b)$$

$$P_{y7} = \frac{-W}{2} \int_{-1}^{1} \frac{1}{4}(1 + \xi\xi_i)(1 + \eta\eta_i)(\xi\xi_i + \eta\eta_i - 1)\,d\xi \qquad (3.42c)$$

In the above equation $\eta = 1$ for all the nodes along the top edge, and therefore, for node 5 $\xi_i = 1$, $\eta_i = 1$, for node 6 $\xi_i = 0$, $\eta_i = 1$, and for node 7 $\xi_i = -1$, $\eta_i = 1$. Substituting for ξ_i, η_i and η in Eqn.(3.42) gives:

$$P_{y5} = \frac{-W}{2} \int_{-1}^{1} \frac{1}{4}(1 + \xi)(2)(\xi + 1 - 1)\,d\xi$$

$$P_{y6} = \frac{-W}{2} \int_{-1}^{1} \frac{1}{2}(2)(1 - \xi^2)\,d\xi \qquad (3.43)$$

$$P_{y7} = \frac{-W}{2} \int_{-1}^{1} \frac{1}{2}(1 - \xi)(2)(-\xi + 1 - 1)\,d\xi$$

Evaluating these integrals gives:

$$P_{y5} = -W/6$$

$$P_{y6} = -2W/3$$

$$P_{y7} = -W/6$$

Hence a uniformly distributed load along an edge is not distributed in the intuitive ratio of $\frac{1}{4} : \frac{1}{2} : \frac{1}{4}$, but in the ratio $\frac{1}{6} : \frac{2}{3} : \frac{1}{6}$ (Fig.3.7(b)).

If the same procedure is used to determine the equivalent nodal force for gravity load (downwards negative) on the rectangular element (Fig. 3.8(a)), then by virtue of Eqn.(3.41),

$$\{P\} = \int_{-1}^{1}\int_{-1}^{1} m[N]^{T}\begin{Bmatrix} 0 \\ -g \end{Bmatrix}\det[J]d\xi d\eta \tag{3.44}$$

This implies that all the equivalent nodal forces act in the y direction, and therefore Eqn.(3.44) can be expanded into

$$\begin{Bmatrix} P_{y1} \\ P_{y2} \\ \vdots \\ \vdots \\ P_{y8} \end{Bmatrix} = -\int_{-1}^{1}\int_{-1}^{1} mg\begin{Bmatrix} N_1 \\ N_2 \\ \vdots \\ \vdots \\ N_8 \end{Bmatrix}\det[J]d\xi d\eta$$

For a corner node i the integral becomes

$$P_{yi} = -\int_{-1}^{1}\int_{-1}^{1} mg\frac{1}{4}(1 + \xi\xi_i)(1 + \eta\eta_i)(\xi\xi_i + \eta\eta_i - 1)\det[J]d\xi d\eta$$

where the values of ξ_i and η_i are determined by the location of the corner node considered.

For a rectangular element, the Jacobian matrix is very simple, and

$$\det[J] = \begin{bmatrix} a & 0 \\ 0 & b \end{bmatrix} = ab$$

Hence

$$P_{yi} = \frac{-mgab}{4}\left[\int_{-1}^{1}\int_{-1}^{1}(1 + \xi\xi_i)(1 + \eta\eta_i)\xi\xi_i d\xi d\eta \right.$$

$$\left. + \int_{-1}^{1}\int_{-1}^{1}(1 + \xi\xi_i)(1 + \eta\eta_i)\eta\eta_i d\xi d\eta - \int_{-1}^{1}\int_{-1}^{1}(1 + \xi\xi_i)(1 + \eta\eta_i)d\xi d\eta\right]$$

Writing $mg(2a)(2b) = W$, which is the total weight of the element,

$$P_{yi} = \frac{-W}{16}\left[I_1 + I_2 + I_3\right]$$

where

$$I_1 = \int_{-1}^{1}\int_{-1}^{1}(1 + \xi\xi_i)(1 + \eta\eta_i)\xi\xi_i \, d\xi d\eta$$

$$= \int_{-1}^{1}(1 + \xi\xi_i)\xi\xi_i \, d\xi\left[\eta + \frac{\eta^2\eta_i}{2}\right]_{-1}^{1}$$

$$= 2\xi_i\left[\left(\frac{\xi^2}{2} + \frac{\xi^3}{3}\xi_i\right)\right]_{-1}^{1}$$

$$= \frac{4}{3}\xi_i^2$$

$$= \frac{4}{3}$$

since $\xi_i = \pm 1$ for all corner nodes.

$$I_2 = \int_{-1}^{1}\int_{-1}^{1}(1 + \xi\xi_i)(1 + \eta\eta_i)\eta\eta_i \, d\xi d\eta$$

$$= \frac{4}{3}\eta_i^2$$

$$= \frac{4}{3}$$

since $\eta_i = \pm 1$ for all corner nodes.

Finally,

$$I_3 = -\int_{-1}^{1}\int_{-1}^{1}(1 + \xi\xi_i)(1 + \eta\eta_i) \, d\xi d\eta$$

$$= -\int_{-1}^{1}(1 + \xi\xi_i) \, d\xi\int_{-1}^{1}(1 + \eta\eta_i) \, d\eta$$

$$= -\left[\xi + \frac{\xi^2\xi_i}{2}\right]_{-1}^{1} * \left[\eta + \frac{\eta^2\eta_i}{2}\right]_{-1}^{1}$$

$$= 4.$$

Therefore, the equivalent nodal forces for gravity loading at a corner node i can be given as

$$P_{yi} = \frac{-W}{16}\left[I_1 + I_2 + I_3\right]$$

$$= \frac{-W}{16}\left[\frac{4}{3} + \frac{4}{3} - 4\right]$$

$$= \frac{+W}{12}$$

For a midside node i with $\xi_i = 0$, the integral is

$$P_{yi} = -\int_{-1}^{1}\int_{-1}^{1} m\frac{1}{2}(1 - \xi^2)(1 + \eta\eta_i)\det\left[J\right]d\xi d\eta$$

$$= \frac{-W}{8}\int_{-1}^{1}(1 + \eta\eta_i)d\eta\int_{-1}^{1}(1 - \xi^2)d\xi$$

$$= \frac{-W}{8}\left[\eta + \frac{\eta^2}{2}\eta_i\right]_{-1}^{1} \times \left[\xi - \frac{\xi^3}{3}\right]_{-1}^{1}$$

$$= \frac{-W}{8}[2]\left[1 - \frac{1}{3} + 1 - \frac{1}{3}\right]$$

$$= \frac{-W}{3}$$

For a midside node i with $\eta_i = 0$, the integral is

$$P_{yi} = -\int_{-1}^{1}\int_{-1}^{1} m\frac{1}{2}(1 - \eta^2)(1 + \xi\xi_i)\det\left[J\right]d\xi d\eta = \frac{-W}{3}$$

Hence for gravity loading the equivalent nodal forces are as shown in Fig.3.8(b), and are thus very different from the values of $-W/12$ and $-W/6$, which would have been assigned intuitively to the corner and midside nodes respectively. The latter type of assigned loading will in fact yield very much poorer results in a finite element analysis using quadratic isoparametric elements.

References

1. O.C. Zienkiewicz, B.M. Irons, J. Ergatoudis, S. Ahmad, and F.C. Scott. Isoparametric and associated element family for two- and three-dimensional analysis. In *Finite Element Methods in Stress Analysis* (I. Holland and K. Bell, eds). New York, Tapir, 1969.
2. Z. Kopal. *Numerical Analysis*, 2nd ed. London, Chapman & Hall, 1961.
3. O.C. Zienkiewicz. *Finite Element Method in Engineering Science*. New York, McGraw-Hill Book Co., 1971.
4. K.J. Bathe and E.L. Wilson. *Numerical Methods in Finite Element Analysis*. Englewood Cliffs, N.J., Prentice-Hall, 1976.
5. K.A. Gupta and B. Mohraz. A method of computing numerically integrated stiffness matrices. *International Journal for Numerical Methods in Engineering,* vol.5, no.1, 1972, pp.83-89.

4 The Front Solver

4.1 Introduction

In the simple program presented in Chapter 2 a full matrix solution and
an in-core band solver were used for solving the set of simultaneous
equations which was formed by assembling the stiffness matrices of all
the elements, and equating the internal forces corresponding to the
displacements to the external loads. Such an approach naturally would
severely restrict the size of the problem that can be tackled, even on
the biggest computer available.

To overcome this limitation it is necessary either to store the
structural stiffness matrix and load matrix on disc files and transfer
to core memory when needed, or to carry out at any one time a simultaneous assembly and reduction on part of the structural stiffness matrix,
in which case no intermediate storage is needed. Two suitable subroutines which will be described here are the out-of-core band solver
and the front solver.

The in-core band solver was described in Section 2.9 and the mathem-
atical details will not be repeated here. For an out-of-core band
solver the diagonal coefficients are no longer shifted to the left as
shown in Fig.2.10. Therefore instead of storing the whole matrix in a
rectangular array, only a part of the equations are stored in a one-
dimensional array, as shown in Fig.4.1(b). Furthermore, to use the
first equation for the reductions only the coefficients of that equation
need to be fully assembled and, furthermore, only the area to the left
of the broken line are involved in the process. This implies that only
the elements connected to node 1 have to be assembled before the
equation (or equations) for node 1 is used for the reduction and so on
for other nodes. Therefore the storage required at all times is $\frac{1}{2}n^2$,
in which n is the half bandwidth of the stiffness matrix.

The out-of-core solver described above can be used to solve problems
involving a large number of degrees of freedom, but it is still inefficient in core storage and computation in that all the zero coefficients
inside the band are stored. For most problems the stiffness matrices
are very sparse and even inside the band the zero coefficients very
often far outnumber the non-zero coefficients.

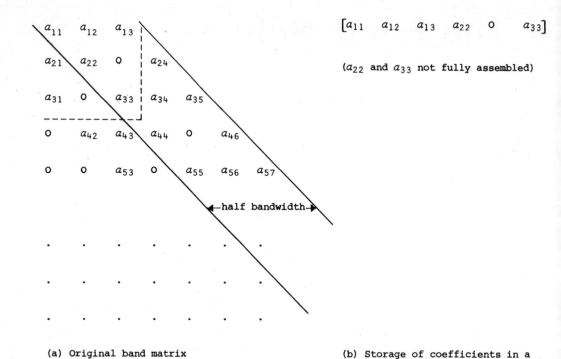

(a) Original band matrix

(b) Storage of coefficients in a vector

Fig.4.1 Out-of-core band solver storage scheme

A more advanced problem which takes advantage of this property is the front solver which will be discussed fully in this chapter.

The front solver is never less efficient than the band solver and for the majority of cases is very much more efficient, especially for problems in which elements with midside nodes are used. The front solver does not store as many zero coefficients inside the band, since unlike the band solver in which the nodal numbers determine the storage locations, it allocates storage in the order in which the nodes are presented for assembly. Thus for the front solver node numbering is irrelevant and it is the element numbering that is all important. This is of great significance when a re-analysis has to be made with local refinement of the finite element mesh.

4.2 The front solver [1]

The operations of a front solver can be split up into three logical parts:
(i) prefront;
(ii) reduction and pre-constraints;
(iii) backsubstitution and post-constraints.

The pre-constraints and post-constraints impose specified displacements on the nodal variables and because these subroutines are self-contained within themselves, they will be omitted for the time being in the small problem used here to demonstrate the operation of the front solver.

Figure 4.2 represents a solid structure, modelled by three elements and constrained against moving both horizontally and vertically at nodes 4 and 2 and loaded by a single vertical load W at node 11 as shown.

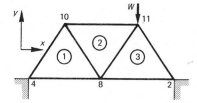

Fig.4.2 An example problem

The element definitions for the structure are:

Element number	Node i	Node j	Node m
1	4	8	10
2	8	11	10
3	2	11	8

Some of the nodal numbers have been missed out purposely to demonstrate the superiority of the front solver over the band solver which cannot cope with such a situation.

4.2.1 PREFRONT

This subroutine determines the nodal destination vector and the maximum front width, which is the equivalent to the half bandwidth for the band solver. The nodal destination vector contains the locations allocated to each node and it is from this vector that the element destination vectors are eventually determined.

The first step is to pin-point the position in which a node number appears for the last time in the element definition list by inserting a negative sign in front of the node number in the appropriate element. This is required because these nodal displacements can be eliminated at this stage as shown below. Thus the element definition vectors for this simple problem would finally look as follows:

$$\begin{bmatrix} -4 & 8 & 10 \end{bmatrix}$$
$$\begin{bmatrix} 8 & 11 & -10 \end{bmatrix}$$
$$\begin{bmatrix} -2 & -11 & -8 \end{bmatrix}$$

It is now possible to determine the nodal and element destination vectors. At the start the displacement vector is completely empty and unallocated, and therefore the element destination vector of the first element is simply $\begin{bmatrix} 1 & 2 & 3 \end{bmatrix}$ and this means that the first node, node 4, is allocated to the first unallocated location in the displacement vector, i.e. location 1, and node 8 and node 10 the next unallocated locations, i.e. location 2 and location 3 respectively. The fourth, eighth and tenth coefficients of the nodal destination vector are therefore set to 1, 2 and 3 respectively.

The displacement vector at this stage contains $\begin{bmatrix} \delta_4 & \delta_8 & \delta_{10} \end{bmatrix}^T$.

The negative sign in front of node 4 in the definition vector means that all equations relating to δ_4 have now been fully assembled and hence δ_4 can be eliminated from the equation system. The equations used to effect this elimination are then removed to disc storage and the displacement vector after this reduction contains $[0 \quad \delta_8 \quad \delta_{10}]^T$.

Element 2 with element definition vector $[8, \quad 11, \quad -10]$ is the next element presented for assembly. Node 8 and node 10 have already been allocated and since node 11 has not appeared previously it is allocated to the first unallocated location which in this instance is location 1. The eleventh coefficient of the nodal destination vector is therefore set to 1. Hence after assembly of element 2 the displacement vector contains $[\delta_{11} \quad \delta_8 \quad \delta_{10}]^T$ and from the nodal destination vector the element destination vector can be determined as $[2, \quad 1, \quad 3]$.

The equations relating to node 10 have now been fully assembled and after reduction and removal to disc storage the displacement vector contains $[\delta_{11} \quad \delta_8 \quad 0]^T$.

Finally, the third element with element definition vector $[-2, \quad -11, \quad -8]$ is presented for assembly. Node 2 is allocated to location 3 and the second coefficient of the nodal destination vector is set to 3. The displacement vector then contains $[\delta_{11} \quad \delta_8 \quad \delta_2]^T$ and the element destination vector is $[3, \quad 1, \quad 2]$.

The equations are used in turn in the reduction process since element 3 is the last element. The sequence is indicated by the element definition vector from left to right and after reduction of the equations corresponding to δ_2 and then δ_{11} the remaining equations will only have coefficients relating to δ_8. This δ_8 can be determined and used as a starting point for the backsubstitution.

It should be mentioned that, as an example, δ_8 refers to all the components of the displacement of node 8, hence for the present problem it will include δ_{x8} and δ_{y8}. Similarly the stiffness coefficients shown as k_{ij} in the following section stand in the present problem for 2×2 submatrices.

In conclusion, after the pre-front subroutine the nodal destination vector contains

$$[0 \quad 3 \quad 0 \quad 1 \quad 0 \quad 0 \quad 0 \quad 2 \quad 0 \quad 3 \quad 1]$$

The zero coefficients indicate node numbers not used in the problem.

Subsequently, from the nodal destination vector subroutine, STIFN sets up the element destination vectors as

$$[1 \quad 2 \quad 3]$$
$$[2 \quad 1 \quad 3]$$
$$[3 \quad 1 \quad 2]$$

The maximum nodal front width is found to be 3, which is the highest number appearing in the nodal destination vector. The actual front width of the problem, defining the size of matrix required is obviously the maximum nodal front width multiplied by the number of variables (displacements) per node.

Subroutine PREFNT

```
C      INSERT COMMON BLOCK


       DO 1 I=1,MAXNW
       NW(I)=0
     1 CONTINUE
       DO 10 NEL=1,MAXNEL
       DO 5 I=1,NNODZ
       NIC=LDEF(NEL,I)
       LDEST(NIC)=NEL
     5 CONTINUE
    10 CONTINUE
       DO 20 NIC=1,MAXNOD
       NEL=LDEST(NIC)
       IF(NEL.EQ.0) GO TO 20
       DO 15 I=1,NNODZ
       IF(LDEF(NEL,I).NE.NIC) GO TO 15
       LDEF(NEL,I)=-NIC
       LDEST(NIC)=0
       GO TO 20
    15 CONTINUE
    20 CONTINUE
       DO 100 NEL=1,MAXNEL
       DO 50 I=1,NNODZ
       NIC=IABS(LDEF(NEL,I))
       IF(LDEST(NIC).NE.0) GO TO 50
       DO 30 J=1,MAXNW
       IF(NW(J).NE.0) GO TO 30
       LDEST(NIC)=J
       NW(J)=NIC
       IF(MAXFW.LT.J) MAXFW=J
       GO TO 50
    30 CONTINUE
       WRITE(6,1000)
       STOP
    50 CONTINUE
       DO 70 I=1,NNODZ
       NIC=LDEF(NEL,I)
       IF(NIC.GT.0) GO TO 70
       N1=LDEST(-NIC)
       NW(N1)=0
    70 CONTINUE
   100 CONTINUE
       MAXFW=MAXFW*NVABZ
       WRITE(6,1001) MAXFW
       RETURN
  1000 FORMAT(62H1MAXIMUM FRONT WIDTH DURING PREFNT EXCEEDS LENGTH OF NW
      1VECTOR)
  1001 FORMAT(//////,5X,19HMAXIMUM FRONT WIDTH,2X,I3)
       END
```

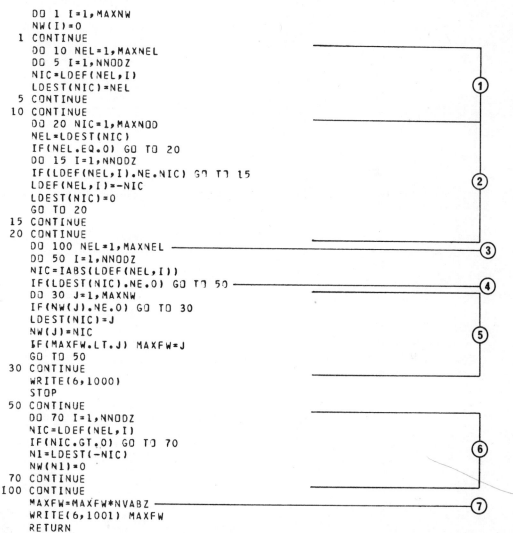

(1) Determine the element in which each node appears last.
(2) Set last appearance of node in the element definition list to a negative number.
(3) Loop 100 sets up the nodal destination vector.
(4) Checking for prior allocation of a node.
(5) Determining the first unallocated column and updating the maximum nodal front width.
(6) The last appearance of a node releases the column for reallocation.
(7) Maximum front width.

4.2.2 ASSEMBLY AND REDUCTION

The reduction procedure in a front solver is virtually the same as that in a band solver, except that the equation being used to reduce the other equations within the front area may occur anywhere in that area and is very often not the top equation.

To continue with the example problem, we have now the element destination vectors determined and the maximum front width calculated as equal to 3. The first element stiffness matrix is now calculated according to Eqn.(3.28) and is of the form

$$
\begin{bmatrix}
k_{ii}^1 & k_{ij}^1 & k_{im}^1 \\
 & k_{jj}^1 & k_{jm}^1 \\
\text{symmetric} & & k_{mm}^1
\end{bmatrix}
\begin{Bmatrix} \delta_4 \\ \delta_8 \\ \delta_{10} \end{Bmatrix}
\left\{ \begin{matrix} \end{matrix} \right\}
\begin{Bmatrix} p_4^1 \\ p_8^1 \\ p_{10}^1 \end{Bmatrix}
\tag{4.1}
$$

where for the present two-dimensional problem all of the ks are 2×2 submatrices.

Because there are no edge loads or body forces we have $p_4^1 = p_8^1 = p_{10}^1 = 0$. However for the sake of generality p_4^1, p_8^1 and p_{10}^1 will be retained. Note that the equations are actually stored in one-dimensional arrays: one for the stiffness matrix and one for the load vector.

$$(k_{ii}^1 \quad k_{ij}^1 \quad k_{im}^1 \quad k_{jj}^1 \quad k_{jm}^1 \quad k_{mm}^1)$$ Structural stiffness vector;

$$(p_4^1 \quad p_8^1 \quad p_{10}^1)$$ Load vector.

The order of storing the actual coefficients of the stiffness matrix is most easily seen by referring to the intermediate output from the program in Chapter 5.

Using the element destination vector $\begin{bmatrix} 1, & 2, & 3 \end{bmatrix}$ the element stiffness is assembled (see Section 2.6) into the structural stiffness matrix which appears as shown in Eqn.(4.1) at this stage.

From the element definition vector $\begin{bmatrix} -4, & 8, & 10 \end{bmatrix}$, it is seen that node 4 has appeared for the last time. The set of equations associated with node 4 can now be used to eliminate δ_4 from the equation system. Following the Gaussian elimination procedure described in Section 2.9 and transferring the coefficients $(k_{ii}, k_{ij}, k_{im}, p_4^1)$ to disc storage we have

$$
\begin{bmatrix}
0 & 0 & 0 \\
 & a_{jj}^* & a_{jm}^* \\
\text{symmetric} & & a_{mm}^*
\end{bmatrix}
\begin{Bmatrix} 0 \\ \delta_8 \\ \delta_{10} \end{Bmatrix}
=
\begin{Bmatrix} 0 \\ p_8^{1*} \\ p_{10}^{1*} \end{Bmatrix}
\tag{4.2}
$$

where a_{jj}^* was substituted for the reduced coefficient k_{jj}^{1*}, etc., to preserve similarity with the change to b_{jm}, c_{jm} in subsequent steps (*see* Eqns.(4.5) and (4.7b)).

Although the transferred coefficients have been written as k_{ii} etc. for neatness of presentation, because the program considers each equation separately and not the set of equations relating to δ_4 then although the equation relating to δ_{4x} will be transferred unaltered the equation relating to δ_{4y} will have been reduced to eliminate δ_{4x}, hence the transferred equation $(k_{ii}, k_{ij}, k_{im}, p_4^\dagger)$ implies that the pivotal submatrix k_{ii} has been triangularised and the other coefficients reduced accordingly.

The stiffness matrix of the second element is now calculated.

$$
\begin{array}{c}
\\
2\\
1\\
3
\end{array}
\begin{array}{ccc}
2 & 1 & 3
\end{array}
\begin{bmatrix}
k_{ii}^2 & k_{ij}^2 & k_{im}^2 \\
 & k_{jj}^2 & k_{jm}^2 \\
\text{symmetric} & & k_{mm}^2
\end{bmatrix}
\begin{Bmatrix}
\delta_8 \\ \delta_{11} \\ \delta_{10}
\end{Bmatrix}
=
\begin{Bmatrix}
p_8^2 \\ p_{11}^2 \\ p_{10}^2
\end{Bmatrix}
\tag{4.3}
$$

This time there is one non-zero load coefficient since $p_{11}^2 = \begin{vmatrix} 0 \\ W \end{vmatrix}$. The assembly follows on the usual way and the structural stiffness is

$$
\begin{bmatrix}
k_{jj}^2 & k_{ij}^2 & k_{jm}^2 \\
 & a_{jj}^* + k_{ii}^2 & a_{jm}^* + k_{im}^2 \\
\text{symmetric} & & a_{mm}^* + k_{mm}^2
\end{bmatrix}
\begin{Bmatrix}
\delta_{11} \\ \delta_8 \\ \delta_{10}
\end{Bmatrix}
=
\begin{Bmatrix}
p_{11}^2 \\ p_8^{1*} + p_8^2 \\ p_{10}^{1*} + p_{10}^2
\end{Bmatrix}
\tag{4.4a}
$$

or

$$
\begin{bmatrix}
b_{ii} & b_{ij} & b_{im} \\
 & b_{jj} & b_{jm} \\
\text{symmetric} & & b_{mm}
\end{bmatrix}
\begin{Bmatrix}
\delta_{11} \\ \delta_8 \\ \delta_{10}
\end{Bmatrix}
=
\begin{Bmatrix}
p_{11}^2 \\ p_8^{1*} + p_8^2 \\ p_{10}^{1*} + p_{10}^2
\end{Bmatrix}
\tag{4.4b}
$$

Note that Eqn.(4.4b) has the same form as Eqn.(4.1) and that in general k_{ij}^2 in Eqn.(4.4a), being usually of submatrix form, should be $[k_{ij}^2]^T$, but the superscript T has been omitted here for ease of presentation. The element definition vector $[8, \quad 11, \quad -10]$ shows that node 10 can be used for reduction and consequently after removing for storage the appropriate coefficients

$$
\begin{bmatrix}
b_{ii}^* & b_{ij}^* & 0 \\
 & b_{jj}^* & 0 \\
 & & 0
\end{bmatrix}
\begin{Bmatrix}
\delta_{11} \\ \delta_8 \\ 0
\end{Bmatrix}
=
\begin{Bmatrix}
p_{11}^{2*} \\ (p_8^{1*} + p_8^2)^* \\ 0
\end{Bmatrix}
\tag{4.5}
$$

The coefficients which have been removed for disc storage are

$(b_{im}, b_{jm}, b_{mm}, (p_{10}^{1*} + p_{10}^2))$.

The third element is now processed in the same way and the element stiffness matrix is

$$
\begin{array}{c} 3 \\ 1 \\ 2 \end{array}
\begin{bmatrix} k_{ii}^3 & k_{ij}^3 & k_{im}^3 \\ & k_{jj}^3 & k_{jm}^3 \\ \text{sym} & & k_{mm}^3 \end{bmatrix}
\begin{Bmatrix} \delta_2 \\ \delta_{11} \\ \delta_8 \end{Bmatrix}
=
\begin{Bmatrix} p_2^3 \\ p_{11}^3 \\ p_8^3 \end{Bmatrix}
\tag{4.6}
$$

(with column labels 3, 1, 2 above the matrix)

All the load coefficients are zero, including p_{11}^3, since the load W at node 11 was assembled previously with element 2.

The element stiffness matrix is assembled into the structural stiffness matrix in accordance with the element destination vector and the structural stiffness matrix is

$$
\begin{bmatrix} b_{ii}^* + k_{jj}^3 & b_{ij}^* + k_{jm}^3 & k_{ij}^3 \\ & b_{jj}^* + k_{mm}^3 & k_{im}^3 \\ \text{symmetric} & & k_{ii}^3 \end{bmatrix}
\begin{Bmatrix} \delta_{11} \\ \delta_8 \\ \delta_2 \end{Bmatrix}
=
\begin{Bmatrix} p_{11}^{2*} + p_{11}^3 \\ (p_8^{1*} + p_8^2)^* + p_8^3 \\ p_2^3 \end{Bmatrix}
\tag{4.7a}
$$

or

$$
\begin{bmatrix} c_{ii} & c_{ij} & c_{im} \\ & c_{jj} & c_{jm} \\ & & c_{mm} \end{bmatrix}
\begin{Bmatrix} \delta_{11} \\ \delta_8 \\ \delta_2 \end{Bmatrix}
=
\begin{Bmatrix} p_{11}^* \\ p_8^* \\ p_2^* \end{Bmatrix}
\tag{4.7b}
$$

The element definition vector is $[-2, \ -11, \ -8]$ and therefore δ_2 is eliminated first, δ_{11} next, finally ending with equation(s) containing δ_8 only.

Thus the first step yields

$$
\begin{bmatrix} c_{ii}^* & c_{ij}^* & 0 \\ & c_{jj}^* & 0 \\ & & 0 \end{bmatrix}
\begin{Bmatrix} \delta_{11} \\ \delta_8 \\ 0 \end{Bmatrix}
=
\begin{Bmatrix} p_{11}^{**} \\ p_8^{**} \\ 0 \end{Bmatrix}
\tag{4.8}
$$

after $(c_{im}, c_{jm}, c_{mm}, p_2^*)$ were removed to disc storage.

The second step yields

$$
\begin{bmatrix} 0 & 0 & 0 \\ & c_{jj}^{**} & 0 \\ & & 0 \end{bmatrix}
\begin{Bmatrix} 0 \\ \delta_8 \\ 0 \end{Bmatrix}
=
\begin{Bmatrix} 0 \\ p_8^{***} \\ 0 \end{Bmatrix}
\tag{4.9}
$$

and the coefficients removed to disc storage were $(c_{ii}^*, c_{ij}^*, 0, p_{11}^{**})$.

The final step involves the transfer of the remaining equation(s) to disc storage (0, c_{jj}^{**}, 0, p_8^{***}). This then locates all the reduced equations in one position, the disc storage, to enable the use of a simple and repetitive backsubstitution procedure.

In the example problem only three sets of equations were dealt with at each step, each 'set' consisting of two equations corresponding to the two components, δ_x and δ_y, for each displacement.

For a bigger structural stiffness matrix, say 6 × 6 with the fourth equation being fully assembled, the reduced form would be

$$\begin{bmatrix} k_{11} & k_{12} & k_{13} & 0 & k_{15} & k_{16} \\ & k_{22} & k_{23} & 0 & k_{25} & k_{26} \\ & & k_{33} & 0 & k_{35} & k_{36} \\ & & & 0 & 0 & 0 \\ & & & & k_{55} & k_{56} \\ \text{symmetric} & & & & & k_{66} \end{bmatrix} \qquad (4.10)$$

and the coefficients removed to core storage would be (k_{14}, k_{24}, k_{34}, k_{44}, k_{45}, k_{46}, p_4).

Subroutine ASMBLE

```
C    INSERT COMMON BLOCK


     N1=0
     DO 24 INOD=1,NNODZ
     IDES=NELDES(INOD)
     I1=(IDES-1)*NVABZ                                    (1)
     I2=(INOD-1)*NVABZ                                    (2)
     DO 3 I=1,NVABZ
     SRS(I1+I)=SRS(I1+I)+RS(I2+I)                         (3)
   3 CONTINUE
     DO 23 JNOD=INOD,NNODZ
     JDES=NELDES(JNOD)
     I3=(JDES-1)*NVABZ                                    (4)
     I4=(JNOD-1)*NVABZ                                    (5)
     DO 22 I=1,NVABZ
     ISS=I1+I
     NW(ISS)=1                                            (6)
     IS=I2+I
     DO 20 J=1,NVABZ
     JSS=I3+J
     IF(IDES.GT.JDES) GO TO 19                            (7)
     JS=I4+J
     IF(IS.GT.JS) GO TO 20
     N1=N1+1                                              (8)
     LOC=JW(ISS)+JSS-ISS                                  (9)
     SS(LOC)=SS(LOC)+S(N1)                                (10)
     GO TO 20
  19 N1=N1+1                                              (8)
     LOC=JW(JSS)+ISS-JSS                                  (9)
     SS(LOC)=SS(LOC)+S(N1)                                (11)
```

```
   20 CONTINUE
   22 CONTINUE
   23 CONTINUE
   24 CONTINUE
      RETURN
      END
```

(1)   Vertical (I) destination in the structural stiffness matrix.
(2)   Vertical (I) location of the next element coefficients to be assembled.
(3)   Assembling the local vector.
(4),(5) As for (1) and (2), except horizontal (J) components.
(6)   Setting corresponding location in heading vector to non-zero.
(7)   If destination in structural stiffness is in the lower triangle then the transformed location is used.
(8)   Nl is the location in the element stiffness vector of the next coefficient to be assembled.
(9)   LOC is the location in the structural stiffness vector into which the element stiffness coefficient is to be assembled.
(10)   Assembly.
(11)   Assembly in the transformed location.

Subroutine FRONT

```
C    INSERT COMMON BLOCK


      REWIND 2
      REWIND 4
      DO 1 I=1,MAXFW
      NW(I)=0
    1 CONTINUE
      JW(1)=1
      DO 2 I=2,MAXFW
      JW(I)=JW(I-1)+MAXFW+2-I
    2 CONTINUE
      MZM=0
      NZN=0
      DO 42 NEL=1,MAXNEL
      READ(4) S,RS,NELDEF,NELDES
      CALL ASMBLE
      NSNW=1
      NFNW=MAXEW
      DO 40 NOD=1,NNODZ
      NIC=-NELDEF(NOD)
      IF(NIC.LE.0) GO TO 40
      LIN=NELDES(NOD)
      CALL PRECON
      DO 38 NODV=1,NVABZ
    9 IF(NW(NSNW).NE.0) GO TO 10
      NSNW=NSNW+1
      GO TO 9
   10 IF(NW(NFNW).NE.0) GO TO 11
      NFNW=NFNW-1
      GO TO 10
   11 LIV=(LIN-1)*NVABZ+NODV
      IF((NZN-NSNW+NFNW+1).LT.MAXRED) GO TO 12
      WRITE(2)  MZM,NZN,REQ,LRED
      MZM=0
      NZN=0
   12 N1=JW(LIV)
      PIVOT=SS(N1)
```

(1) (2) (3) (4) (5) (6) (7)

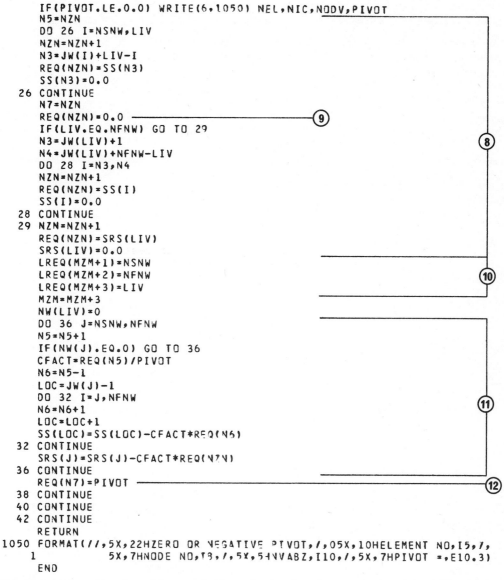

```
        IF(PIVOT.LE.0.0) WRITE(6,1050) NEL,NIC,NODV,PIVOT
        N5=NZN
        DO 26 I=NSNW,LIV
        NZN=NZN+1
        N3=JW(I)+LIV-I
        REQ(NZN)=SS(N3)
        SS(N3)=0.0
     26 CONTINUE
        N7=NZN
        REQ(NZN)=0.0  ─────────────────────────── ⑨
        IF(LIV.EQ.NFNW) GO TO 29
        N3=JW(LIV)+1
        N4=JW(LIV)+NFNW-LIV
        DO 28 I=N3,N4
        NZN=NZN+1
        REQ(NZN)=SS(I)
        SS(I)=0.0
     28 CONTINUE
     29 NZN=NZN+1
        REQ(NZN)=SRS(LIV)
        SRS(LIV)=0.0
        LREQ(MZM+1)=NSNW
        LREQ(MZM+2)=NFNW
        LREQ(MZM+3)=LIV
        MZM=MZM+3
        NW(LIV)=0
        DO 36 J=NSNW,NFNW
        N5=N5+1
        IF(NW(J).EQ.0) GO TO 36
        CFACT=REQ(N5)/PIVOT
        N6=N5-1
        LOC=JW(J)-1
        DO 32 I=J,NFNW
        N6=N6+1
        LOC=LOC+1
        SS(LOC)=SS(LOC)-CFACT*REQ(N6)
     32 CONTINUE
        SRS(J)=SRS(J)-CFACT*REQ(NZN)
     36 CONTINUE
        REQ(N7)=PIVOT  ─────────────────────────── ⑫
     38 CONTINUE
     40 CONTINUE
     42 CONTINUE
        RETURN
   1050 FORMAT(//,5X,22HZERO OR NEGATIVE PIVOT,/,05X,10HELEMENT NO,I5,/,
        1          5X,7HNODE NO,I3,/,5X,5HNVABZ,I10,/,5X,7HPIVOT =,E10.3)
        END
```

(1) Determine the locations of the pivots in the structural stiffness vector.
(2) Loop 42 is the element loop.
(3) For each element in turn read the stiffness, load vector, definition vector and destination vector.
(4) Check for the last appearance of each of the element nodes.
(5) Location of the pivotal equations.
(6) Reset the start and finish flags for the heading vector.
(7) If the temporary storage buffer for reduced equations is full, the buffer is transferred to permanent storage.
(8) The next equation to be used for reduction is stored in the temporary storage buffer.
(9) Set the pivotal coefficient in the stored equation to zero.
(10) Store the start and finish flags and the position of the pivot.
(11) Reduce the stiffness and load vectors.
(12) Replace the value of the pivot in the reduced equation.

4.2.3 BACKSUBSTITUTION

The backsubstitution is a more complicated procedure in the front
solver than in the band solver because of the variable location of the
pivotal coefficients in the stored equations. The non-sequential
nature of the displacement vector likewise complicates the procedure.
However, if a reverse procedure to reduction is used to determine the
location of the pivotal coefficient in each of the stored equations and
also the associated nodal number, then a single repetitive system can
be set up to determine the displacements.

The element destination vectors are used to indicate the position of
the pivots and the element definition vectors to determine the displacement associated with a particular pivot.

The data which is stored after the completion of reduction is listed
in Table 4.1.

TABLE 4.1 *Data stored for use in backsubstitution*

ELEM #

Element definition vector	-4	8	10	1
	8	11	-10	2
	-2	-11	-8	3
Element destination vector	1	2	3	1
	2	1	3	2
	3	1	2	3
Reduced equations	$(k_{ii}^1, k_{ij}^1, k_{im}^1, p_4^1)$			
	$(b_{im}, b_{jm}, b_{mm}, (p_{10}^{1*} + p_{10}^2))$			
	$(c_{im}, c_{jm}, c_{mm}, p_2^*)$			
	$(c_{ii}^*, c_{ij}^*, 0, p_{11}^{**})$			
	$(0, c_{jj}^*, 0, p_8^{***})$			

The backsubstitution procedure must be made automatic and repetitive
so that an identical process is carried out to calculate the value of
each nodal displacement. This programming philosophy saves a great
amount of complicated FORTRAN coding and hence execution time. The
method implemented in the program which is listed at the end of this
section uses a working vector equal in length to the front width. For
the two-dimensional problem this is equal to the nodal front width
multiplied by 2, i.e. 3 × 2 = 6 in the present case.

Initially the vector contains only zeros but later on this changes to
the vector of displacements which are being continuously updated as the
backsubstitution progresses.

The procedure commences by searching the element definition vectors for negative node numbers starting with the last element and working from the right to left of each definition vector. This operation determines the reverse order to which the nodal equations were eliminated. The first negative node number detected is that of node 8, and from Table 4.1 it is seen that the corresponding coefficient in the destination vector is 2, indicating that the second coefficient, c_{jj}^{**}, in the stored equation is the pivot.

As indicated earlier in both the reduction and backsubstitution stages of the program each equation of the set of equations related to a node is considered separately. Considering the last equation $c_{jj}^{**}\delta_8 = p_8^{***}$ in its expanded form it can be represented as

$$\begin{bmatrix} _{11}c_{jj}^{**} & _{12}c_{jj}^{**} \\ 0 & (_{22}c_{jj}^{**})^* \end{bmatrix} \begin{bmatrix} \delta_{8x} \\ \delta_{8y} \end{bmatrix} = \begin{bmatrix} p_{8x}^{***} \\ (p_{8y}^{***})^* \end{bmatrix}$$

The stored equation relating to δ_{8y} is therefore

$$\begin{bmatrix} 0, & 0, & 0, & (_{22}c_{jj}^{**})^*, & 0, & 0, & (p_{8y}^{***})^* \end{bmatrix}$$

To make the process automatic a local variable PIVOT (see computer program) is set equal to the value of the pivotal coefficient which in turn is set equal to zero. Therefore at this instant

PIVOT $= (_{22}c_{jj}^{**})^*$

and the stored equation becomes $(0, 0, 0, 0, 0, 0, (p_{8y}^{***})^*)$. The reason for setting the pivotal coefficient to zero will become clear later.

The standard equation for calculating a displacement can be established as

Nodal displacement =

$$\frac{\text{Load} - \text{(Stored coefficients)(Running vector of displacements)}}{\text{PIVOT}}$$

The above equation is applicable to all nodes, and this includes the last node since

$$\delta_{8y} = \frac{\left((p_{8y}^{***})^* - \begin{bmatrix} 0, & 0, & 0, & 0, & 0, & 0 \end{bmatrix}\begin{bmatrix} 0, & 0, & 0, & 0, & 0, & 0 \end{bmatrix}^T \right)}{(_{22}c_{jj}^{**})^*}$$

is the result that would have been obtained if the last stored equation had been used directly.

The value of δ_{8y} is now stored in the appropriate location in the running vector of displacements as determined by the destination vector and the running vector changes from $(0, 0, 0, 0, 0, 0)$ to $(0, 0, 0, \delta_{8y}, 0, 0)$.

The same approach is used to determine δ_{8x}.

PIVOT = $_{11}c_{jj}^{**}$

and the stored equation is modified into $(0, 0, 0, {}_{12}c_{jj}^{**}, 0, 0, p_{8x}^{***})$.
Hence δ_{8x} is determined by the standard equation,

$$\delta_{8x} = \frac{p_{8x}^{***} - [0, 0, 0, {}_{12}c_{jj}^{**}, 0, 0][0, 0, 0, \delta_{8y}, 0, 0]^T}{{}_{11}c_{jj}^{**}}$$

and the running vector of displacements is updated to
$(0, 0, \delta_{8x}, \delta_{8y}, 0, 0)$.

To avoid continuous repetition the operations to calculate the x and y components of the displacement at a node will be written in terms of the submatrices. The above two operations to calculate δ_{8x} and δ_{8y} could be written as

Stored Pivot = c_{jj}^{**} and

$$\delta_8 = [c_{jj}^{**}]^{-1}(p_8^{***} - [0, 0, 0][0, 0, 0]^T)$$

and the value of δ_8 stored in the running vector changes the running vector to $(0, \delta_8, 0)$.

The search continues through the definition vectors for the next negative node which is node 11 with a destination of 1. Using the same process as for node 8

Stored pivot = c_{ii}^{**}

and the stored equation is modified into $(0, c_{ij}^{*}, 0, p_{11}^{**})$.
Hence δ_{11} is determined by the standard equation

$$\delta_{11} = [c_{ii}^{*}]^{-1}(p_{11}^{**} - [\bullet, c_{ij}^{*}, \bullet][\bullet, \delta_8, \bullet]^T)$$

and the running vector of displacements is updated to $(\delta_{11}, \delta_8, 0)$.

Node 2 is the next negative node and has a destination of 3. Following the same procedure we have

Stored Pivot = c_{mm}

Modified equation = $(c_{im}, c_{jm}, 0, p_2^{*})$

$$\delta_2 = [c_{mm}]^{-1}(p_2^{*} - [c_{im}, c_{jm}, \bullet][\delta_{11}, \delta_8, 0]^T)$$

and the running vector of displacements is now $(\delta_{11}, \delta_8, \delta_2)$.

The next node with a negative node number is node 10 in element 2 with a destination of 3, giving

Stored Pivot = b_{mm}

Modified equation = $(b_{im}, b_{jm}, \bullet, (p_{10}^{1*} + p_{10}^{2}))$

$$\delta_{10} = [b_{mm}]^{-1}\left((p_{10}^{1*} + p_{10}^{2}) - [b_{im}, b_{jm}, \bullet][\delta_{11}, \delta_8, \delta_2]^T\right)$$

The need to set the pivotal coefficient to zero is now seen clearly. Had this not been done the vector multiplication in the above equation would have produced a term containing δ_2 which is not found in the original equation (*see* Eqn.(4.4b)).

The running vector now becomes $(\delta_{11}, \delta_8, \delta_{10})$ and the last node is node 4 from element 1 with a destination of 1. The final step of the backsubstitution procedure ends with

Stored Pivot = k_{ii}^1

Modified equation = $(0, k_{ij}^1, k_{im}^1, p_4^1)$

$$\delta_4 = [k_{ii}^1]^{-1}\left(p_4^1 - [0, k_{ij}^1, k_{im}^1][\delta_{11}, \delta_8, \delta_{10}]^T\right)$$

and the running vector becomes $(\delta_4 \quad \delta_8 \quad \delta_{10})$.

Subroutine BAKSUB

```
C      INSERT COMMON BLOCK

       DO 300 NEL=1,MAXNEL
       MEL=MAXNEL+1-NEL
       NNN=1
       DO 299 I=1,NNODZ
       N1=NNODZ+1-I
       NIC=-LDEF(MEL,N1)
       IF(NIC.LE.0) GO TO 299                       (1)
       DO 230 NODV=1,NVABZ
       IF(NZN.NE.0) GO TO 219
       BACKSPACE 2
       READ(2) MZM,NZN,REQ,LREQ                     (2)
       BACKSPACE 2
   219 MZM=MZM-3
       NSNW=LREQ(MZM+1)
       NFNW=LREQ(MZM+2)                             (3)
       LIV=LREQ(MZM+3)
       GASH=REQ(NZN)                                (4)
       PIVOT=REQ(NZN-NFNW-1+LIV)
       REQ(NZN-NFNW-1+LIV)=0.0                      (5)
       N2=NZN-NFNW-2+NSNW
       NZN=N2
       DO 220 J=NSNW,NFNW
       N2=N2+1
       GASH=GASH-SRS(J)*REQ(N2)                     (6)
   220 CONTINUE
       SRS(LIV)=GASH/PIVOT
       N3=NVABZ+1-NODV
       DISPL(NIC,N3)=GASH/PIVOT                     (7)
   230 CONTINUE
       CALL POSTCN(NNN,MEL)
   299 CONTINUE
   300 CONTINUE
       IF(MAXTNS.NE.0) CALL BTRANS
       RETURN
       END
```

(1) Determine the nodal order of back substitution.
(2) Refill the temporary storage buffer if empty.
(3) Location of the pivot, start and finish flags.
(4) Load coefficient.
(5) Set the pivot to zero.
(6) Calculate the displacement.
(7) Store the calculated displacement.

4.2.4 PRE-CONSTRAINTS AND POST-CONSTRAINTS

The technique for imposing specified displacements for the simple
program in Chapter 2 will also be used in the front solver, and it was
seen that in the simple program all the nodes were 'aware' of a specified
displacement, say δ_n, because the load terms for all the equations were
modified simultaneously. However, for a front solver only a small
number of equations are in core at any one time and equation n is only
fully assembled immediately prior to being used for reduction, which
means that it is not possible to specify δ_n beforehand. Hence although
all the nodes in core at that time and all subsequent nodes brought into
core will be 'aware' of the specified δ_n, nodes eliminated prior to that
time have no such 'awareness' since their load terms have not been
modified.

The operations for imposing specified displacements is therefore split
into two parts. The pre-constraint subroutine PRECON is called during
the reduction process immediately before a node is eliminated. If node
n has a specified displacement then the current load vector is modified
and the pivot is increased by a big spring stiffness, in accordance with
Eqn. (2.39).

The post-constraint subroutine POSTCN is called immediately after the
computation of a nodal displacement during the backsubstitution process.
When it comes across node n with specified displacement δ_n it will
firstly calculate the reaction (see Method 4, Section 2.7) and secondly
replace the small calculated displacement by δ_n in the running displacement vector. In this way the previously eliminated nodes are also made
'aware' of the specified displacement since δ_n will actually be used in
calculating the displacements of these nodes.

Subroutine PRECON

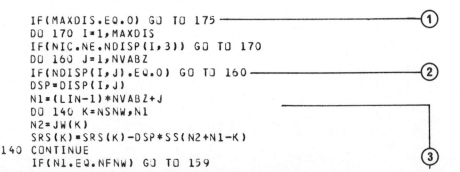

```
C     INSERT COMMON BLOCK


      IF(MAXDIS.EQ.0) GO TO 175 ──────────────────────────────①
      DO 170 I=1,MAXDIS
      IF(NIC.NE.NDISP(I,3)) GO TO 170
      DO 160 J=1,NVABZ
      IF(NDISP(I,J).EQ.0) GO TO 160 ───────────────────────────②
      DSP=DISP(I,J)
      N1=(LIN-1)*NVABZ+J
      DO 140 K=NSNW,N1
      N2=JW(K)
      SRS(K)=SRS(K)-DSP*SS(N2+N1-K)
  140 CONTINUE
      IF(N1.EQ.NFNW) GO TO 159 ────────────────────────────────③
```

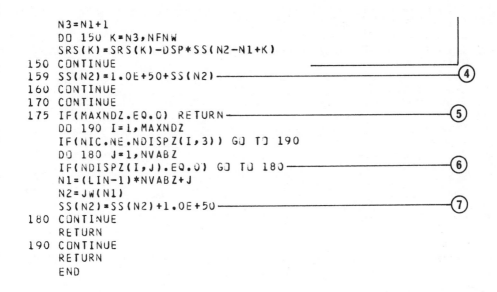

```
      N3=N1+1
      DO 150 K=N3,NFNW
      SRS(K)=SRS(K)-DSP*SS(N2-N1+K)
 150  CONTINUE
 159  SS(N2)=1.0E+50+SS(N2)
 160  CONTINUE
 170  CONTINUE
 175  IF(MAXNDZ.EQ.0) RETURN
      DO 190 I=1,MAXNDZ
      IF(NIC.NE.NDISPZ(I,3)) GO TO 190
      DO 180 J=1,NVABZ
      IF(NDISPZ(I,J).EQ.0) GO TO 180
      N1=(LIN-1)*NVABZ+J
      N2=JW(N1)
      SS(N2)=SS(N2)+1.0E+50
 180  CONTINUE
      RETURN
 190  CONTINUE
      RETURN
      END
```

(1) Check if any of the nodes have specified non-zero displacements.
(2) Determine which variables have been specified.
(3) Subtrace from the load vector the specified displacement multiplied by the associated column of the stiffness matrix.
(4) Add the big spring stiffness to the pivotal coefficient.
(5) Check if any of the nodes have specified earthed displacements.
(6) Determine which variables have been specified.
(7) Add the big spring stiffness to the pivotal coefficient.

Subroutine POSTCN(NNN, MEL)

```
C     INSERT COMMON BLOCK

      DIMENSION RACTN(2)
      DO 232 I=1,NVABZ
      RACTN(I)=0.0
 232  CONTINUE
      IF(MAXDIS.EQ.0) GO TO 251
      DO 250 I=1,MAXDIS
      IF(NIC.NE.NDISP(I,3)) GO TO 250
      DO 240 J=1,NVABZ
      IF(NDISP(I,J).EQ.0) GO TO 240
      DSP=DISP(I,J)
      N1=LIV+J-1
      RACTN(J)=-SRS(N1)*1.0E+50
      SRS(N1)=DSP
      DISPL(NIC,J)=DSP
 240  CONTINUE
      IF(NNN.EQ.1) WRITE(6,1005) MEL
      NNN=0
      WRITE(6,1200) NIC,RACTN
 250  CONTINUE
 251  IF(MAXNDZ.EQ.0) RETURN
```

```
      DO 270 I=1,MAXNDZ
      IF(NIC.NE.NDISPZ(I,3)) GO TO 270
      DO 260 J=1,NVABZ
      IF(NDISPZ(I,J).EQ.0) GO TO 260 ─────────────────── ⑥
      N1=LIV+J-1
      RACTN(J)=-SRS(N1)*1.0E+50 ─────────────────── ⑦
      SRS(N1)=0.0
      DISPL(NIC,J)=0.0 ─────────────────── ⑧
  260 CONTINUE
      IF(NNN.EQ.1) WRITE(6,1005) MEL
      NNN=0
      WRITE(6,1200) NIC,RACTN
  270 CONTINUE
      RETURN
 1005 FORMAT(///,5X,18HREACTIONS ELEMENT ,I3,//,
    1          7X,*NODE*,7X,*1ST COMP*,7X,*2ND COMP*)
 1200 FORMAT(//,6X,I5,3(5X,E10.3))
      END
```

(1) Check if any node has any non-zero displacements.
(2) Determine which variables have been specified.
(3) Calculate the reaction.
(4) Replace the small calculated displacement by the specified displacement.
(5) Check if any node has specified earthed displacements.
(6) Determine which variables have been specified.
(7) Calculate the reaction.
(8) Replace the small calculated displacement by zero.

4.3 Transformed constraints

It is very often possible to reduce the size of the structure to be modelled by using locally oriented displacements. For example the simple problem in Chapter 2 is axisymmetric, and instead of analysing a quarter of the disc it is really only necessary to use a small segment for a complete analysis, as shown in Fig.4.3.

From Fig.4.3(b) it is seen that some of the nodes are now defined with

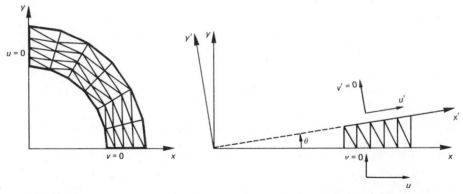

(a) Mesh layout for standard constraints   (b) Mesh layout for transformed constraints

Fig.4.3 Two different mesh layouts for an axisymmetric problem

respect to their local axes, and Eqn.(1.45) which was originally given in global axes must undergo some sort of transformation process. The transformation between two sets of coordinate systems is well known [2] and is given by

$$\{\delta_i\} = \begin{Bmatrix} u_i \\ v_i \end{Bmatrix} = [L]\begin{Bmatrix} u_i' \\ v_i' \end{Bmatrix} = [L]\{\delta_i'\}$$

$$(4.11)$$

$$\{F_i'\} = \begin{Bmatrix} F_{xi}' \\ F_{yi}' \end{Bmatrix} = [L]^T\begin{Bmatrix} F_{xi} \\ F_{yi} \end{Bmatrix} = [L]^T\{F_i\}$$

and

$$[L] = \begin{bmatrix} \cos\theta & -\sin\theta \\ \sin\theta & \cos\theta \end{bmatrix}$$

In the above equations the global variables are non-primed and the local variables are primed, with θ taken as positive as shown by the arrow in Fig.4.3(b).

It is now assumed that the displacements of node i are locally oriented, then Eqn.(1.45) should be rewritten as

$$\begin{Bmatrix} F_i \\ F_{i-1} \\ F_i' \\ F_{i+1} \\ F_{11} \end{Bmatrix} = \begin{bmatrix} I & & & & \\ & I & & & \\ & & L^T & & \\ & & & I & \\ & & & & I \end{bmatrix}\begin{Bmatrix} F_1 \\ F_{i-1} \\ F_i \\ F_{i+1} \\ F_{12} \end{Bmatrix}$$

$$= \begin{bmatrix} I & & & & \\ & I & & & \\ & & L^T & & \\ & & & I & \\ & & & & I \end{bmatrix}[K]\begin{Bmatrix} \delta_1 \\ \delta_{i-1} \\ \delta_i \\ \delta_{i+1} \\ \delta_n \end{Bmatrix}$$

$$= \begin{bmatrix} I & & & & \\ & I & & & \\ & & L^T & & \\ & & & I & \\ & & & & I \end{bmatrix}[K]\begin{bmatrix} I & & & & \\ & I & & & \\ & & L & & \\ & & & I & \\ & & & & I \end{bmatrix}\begin{Bmatrix} \delta_1 \\ \delta_{i-1} \\ \delta_i' \\ \delta_{i+1} \\ \delta_n \end{Bmatrix}$$

$$
= \begin{bmatrix}
K_{11} & \cdots & K_{1,i-1} & K_{1,i}L & K_{1,i+1} & \cdots & K_{1,n} \\
\vdots & & \vdots & \vdots & \vdots & & \vdots \\
K_{i-1,1} & \cdots & K_{i-1,i-1} & K_{i-1,i}L & K_{i-1,i+1} & \cdots & K_{i-1,n} \\
L^T K_{i,1} & \cdots & L^T K_{i,i-1} & L^T K_{i,i}L & L^T K_{i,i+1} & \cdots & L^T K_{i,n} \\
K_{i+1,1} & \cdots & K_{i+1,i-1} & K_{i+1,i}L & K_{i+1,i+1} & \cdots & K_{i+1,n} \\
\vdots & & \vdots & \vdots & \vdots & & \vdots \\
K_{n,1} & \cdots & K_{n,i-1} & K_{n,i}L & K_{n,i+1} & \cdots & K_n
\end{bmatrix}
\begin{Bmatrix}
\delta_1 \\ \vdots \\ \delta_{i-1} \\ \delta_i' \\ \delta_{i+1} \\ \vdots \\ \delta_n
\end{Bmatrix}
\qquad (4.12)
$$

Hence whenever a variable is transformed into a local set of axes, all
that is required in carrying out the transformation is to post-multiply
the appropriate column by $[L]$ and then pre-multiply the appropriate row
by $[L]^T$. This pre- and post-multiplication is carried out by subroutine
TRANSF. After the solution of the equations the local displacements
should be reorientated into the global system before element stresses
are computed from the displacements. The reorientation of local displacements is carried out by the subroutine BTRANS.

Subroutine TRANSF

```
C     INSERT COMMON BLOCK


      DO 500 I=1,MAXTNS
      NIC=NTRANS(I,1)
      DO 400 J=1,NNODZ
      IF(NELDEF(J).NE.NIC) GO TO 400 ──────────────① 
      NTR=NTRANS(I,2) ──────────────────────────② 
      NLOC=0
      IF(J.EQ.1) GO TO 150
      N1=J-1
      DO 100 I1=1,N1 ───────────────────────────③ 
      LOC=NLOC+3+(J-I1-1)*4 ─────────────────────④ 
      DO 50 I2=1,4
      CW(I2)=S(I2+LOC)
   50 CONTINUE
      S(LOC+1)=CW(1)*TRANS(NTR,1)+CW(2)*TRANS(NTR,3)
      S(LOC+2)=CW(1)*TRANS(NTR,2)+CW(2)*TRANS(NTR,4)
      S(LOC+3)=CW(3)*TRANS(NTR,1)+CW(4)*TRANS(NTR,3)   ⑤ 
      S(LOC+4)=CW(3)*TRANS(NTR,2)+CW(4)*TRANS(NTR,4)
      NLOC=NLOC+3+(8-I1)*4
  100 CONTINUE
  150 DO 160 I2=1,3
      CW(I2)=S(I2+NLOC)
  160 CONTINUE
      A2=CW(1)*TRANS(NTR,2)+CW(2)*TRANS(NTR,4)
      A5=CW(2)*TRANS(NTR,2)+CW(3)*TRANS(NTR,4)
      S(NLOC+2)=TRANS(NTR,1)*A2+TRANS(NTR,3)*A5        ⑥ 
      S(NLOC+3)=TRANS(NTR,2)*A2+TRANS(NTR,4)*A5
      S(NLOC+1)=TRANS(NTR,1)*(CW(1)*TRANS(NTR,1)+CW(2)*TRANS(NTR,3))+
     2          TRANS(NTR,3)*(CW(2)*TRANS(NTR,1)+CW(3)*TRANS(NTR,3))
```

```
      NLOC=NLOC+3
      N2=(J-1)*NVABZ
      A1=RS(N2+1)
      A2=RS(N2+2)
      RS(N2+1)=TRANS(NTR,1)*A1+TRANS(NTR,3)*A2
      RS(N2+2)=TRANS(NTR,2)*A1+TRANS(NTR,4)*A2         (7)
      IF(J.EQ.NNODZ) GO TO 500
      N3=J+1
      DO 300 I3=N3,NNODZ                               (8)
      DO 250 I4=1,4
      CW(I4)=S(NLOC+I4)
  250 CONTINUE
      S(NLOC+1)=TRANS(NTR,1)*CW(1)+TRANS(NTR,3)*CW(3)
      S(NLOC+2)=TRANS(NTR,1)*CW(2)+TRANS(NTR,3)*CW(4)
      S(NLOC+3)=TRANS(NTR,2)*CW(1)+TRANS(NTR,4)*CW(3)  (9)
      S(NLOC+4)=TRANS(NTR,2)*CW(2)+TRANS(NTR,4)*CW(4)
      NLOC=NLOC+4
  300 CONTINUE
      GO TO 500
  400 CONTINUE
  500 CONTINUE
      RETURN
      END
```

(1) Check if any of the element nodes have locally defined axes.
(2) NTR is the number of the transformation matrix.
(3) Loop 100 transforms the column.
(4) Location is SS( ) vector of each 2 × 2 submatrix in turn.
(5) Transformation of each 2 × 2 matrix.
(6) Transformation of the upper triangle of the 2 × 2 pivotal matrix.
(7) Transformation of the load vector.
(8) Loop 300 transforms the row.
(9) Transformation of each 2 × 2 matrix along the row.

Subroutine BTRANS

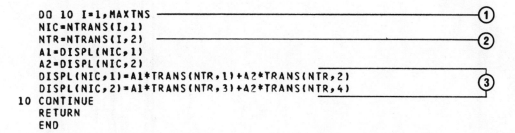

```
C     INSERT COMMON BLOCK

      DO 10 I=1,MAXTNS                               (1)
      NIC=NTRANS(I,1)
      NTR=NTRANS(I,2)                                (2)
      A1=DISPL(NIC,1)
      A2=DISPL(NIC,2)
      DISPL(NIC,1)=A1*TRANS(NTR,1)+A2*TRANS(NTR,2)
      DISPL(NIC,2)=A1*TRANS(NTR,3)+A2*TRANS(NTR,4)   (3)
   10 CONTINUE
      RETURN
      END
```

(1) Loop 10 considers each of the transformed nodes in turn.
(2) NTR is the number of the transformation matrix.
(3) Transformation of the displacements back into the global system.

4.4 Improvements to the efficiency of execution

4.4.1 HEADING VECTOR

During various stages in the reduction process the front area occasionally contains blocks of zero coefficients. It is possible to increase
the efficiency of execution by using an indicator called the heading
vector to identify the locations of these zero blocks so that they can
be completely ignored for computational purposes. It is also possible
via the heading vector to store a variable length equation that does not
contain any zero coefficients which occur either at the beginning or the
end of the equation, so that data transfer between core and disc can be
reduced.

The heading vector contains as many locations as the maximum front
width of the problem. If a column, and hence the corresponding row,
contains only zero coefficients then the associated location in the
heading vector is set to zero. Likewise non-zero coefficients in the
heading vector indicate non-zero coefficients in the associated column
and row.

Consider the frontal area of a structural stiffness matrix, as shown
in Fig.4.4 in which for argument's sake only the crosses represent non-
zero numbers

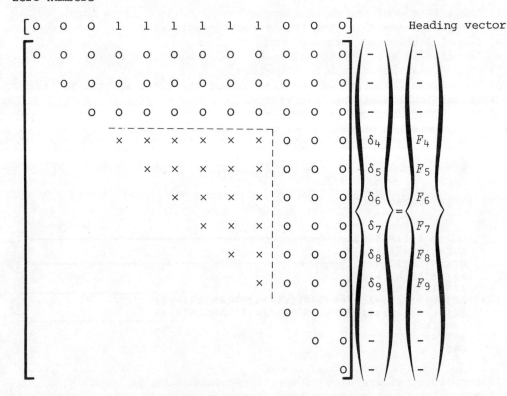

Fig.4.4 Use of the heading vector in a front solver

If the next variable to be eliminated is to be δ_6 then the usual
method would be to store the coefficient relating to node 6

$$(0 \quad 0 \quad 0 \quad k_{46} \quad k_{56} \quad k_{66} \quad k_{67} \quad k_{68} \quad k_{69} \quad 0 \quad 0 \quad 0 \quad F_6)$$

This is wasteful of transfer time as there are six zero coefficients in the equation. If the heading vector is utilized and, indicating the first and last column and row positions with non-zero coefficient are kept updated after each elimination, then only part of the row and column between the start and finish flags need be considered. For the above matrix the start flag NSNW=4 and the finish flag NFNW=9 effectively contract the front area to that enclosed by the broken line. The coefficients for node 6 that are stored are then

$$(k_{46} \quad k_{56} \quad k_{66} \quad k_{67} \quad k_{68} \quad k_{69} \quad F_6) \text{ NSNW NFNW.}$$

The two flags also have to be retained at the end of each equation to indicate how many coefficients have been deleted.

Blocks of zeros occurring in the middle of non-zero coefficients must be stored since the above technique requires storage of all coefficients between the first and last non-zero coefficients of an equation. But by using the heading vector rows containing only zero coefficients can be ignored thus reducing the number of coefficients to be operated upon during the reduction process (see discussion below concerning the elimination of δ_6, δ_4 and δ_3, using the normal front).

The efficiency of such a scheme can be seen by comparing the number of non-zero coefficients in the matrix that would be operated upon when using the heading vector compared to the total number of coefficients in the front area.

The technique of using the heading vector has been implemented into the solution subroutines of the listed program.

4.4.2 LONGEVITY FRONT [3]

The front solver in its present stage of development still occasionally allows columns containing only zero coefficients to occur in the front area among columns containing non-zero coefficients. However, the frequency of such occurrence can be reduced by rearranging the storage of the nodes in the front area according to their 'life span'. Nodes which exist in the front area over the greatest number of elements, the longest lived nodes, are assigned storage locations before nodes with shorter lives.

The easiest means of showing the effect of this longevity is to perform a simple operation count on a two element problem solved firstly by a normal front solver and then by a longevity front.

Consider the two element problem in Fig.4.5. It is assumed that both types of solver have the heading vector facility implemented.

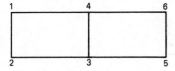

Fig.4.5 Example problem for longevity front

The element definitions are for the normal front

Element 1 $\begin{bmatrix} 4 & -1 & -2 & 3 \end{bmatrix}$

2 $\begin{bmatrix} -5 & -6 & -4 & -3 \end{bmatrix}$

and the element destinations are

Element 1 $\begin{bmatrix} 1 & 2 & 3 & 4 \end{bmatrix}$

2 $\begin{bmatrix} 2 & 3 & 1 & 4 \end{bmatrix}$

The operations required to reduce the upper triangle of the stiffness matrix and the load vector produced by this problem using the normal front are shown below.

First, element 1 is assembled and reduced

$$
\begin{bmatrix}
\times & \times & \times & \times \\
 & \times & \times & \times \\
 & & \times & \times \\
 & & & \times
\end{bmatrix}
\begin{Bmatrix}
\delta_4 \\
\delta_1 \\
\delta_2 \\
\delta_3
\end{Bmatrix}
=
\begin{Bmatrix}
F_4 \\
F_1 \\
F_2 \\
F_3
\end{Bmatrix}
$$

The reduction is carried out in rows so the heading vector will only check on rows not on columns.

(i) δ_1 is eliminated (negative sign for node 1),
    5 sets of coefficients written to store,
    14 sets of coefficients modified (including load terms).

(ii) δ_2 is eliminated (negative sign for node 2),
    5 sets of coefficients written to store,
    10 sets of coefficients modified.

Second, element 2 is assembled,

$$
\begin{bmatrix}
\times & \times & \times & \times \\
 & \times & \times & \times \\
 & & \times & \times \\
 & & & \times
\end{bmatrix}
\begin{Bmatrix}
\delta_4 \\
\delta_5 \\
\delta_6 \\
\delta_3
\end{Bmatrix}
=
\begin{Bmatrix}
F_4 \\
F_5 \\
F_6 \\
F_3
\end{Bmatrix}
$$

(i) δ_5 is eliminated,
    5 sets of coefficients written to store,
    14 sets of coefficients modified.

(ii) δ_6 is eliminated,
    5 sets of coefficients written to store,
    10 sets of coefficients modified (row 2 contains only zeros after reduction in (i) and is ignored completely through the use of the heading vector).

(iii) δ_4 is eliminated,
    5 sets of coefficients written to store,
    7 sets of coefficients modified (rows two and three ignored).

(iv) δ_3 is eliminated,
   2 sets of coefficients written to store,
   2 sets of coefficients modified (rows one, two and three ignored).
Thus we find that the reduction process for the normal front solver requires 84 operations in which 27 sets of coefficients are stored and 57 sets of coefficients modified.
Repeating the operation using the longevity front solver results in a lower operations count as is detailed below. It should be realized that when two nodes have the same element life, the node that occurs farthest towards the right of the definition vector in which both appear *last* is the node with the longer life.
The element definitions are unchanged:

Element 1 $\begin{bmatrix} 4 & -1 & -2 & 3 \end{bmatrix}$

2 $\begin{bmatrix} -5 & -6 & -4 & -3 \end{bmatrix}$

The element destination vectors for the longevity front are different from the normal front and are given as

Element 1 $\begin{bmatrix} 3 & 1 & 2 & 4 \end{bmatrix}$

2 $\begin{bmatrix} 1 & 2 & 3 & 4 \end{bmatrix}$

It is seen that nodes 3 and 4 have the longest life among the six nodes.
   First, element 1 is assembled and reduced

$$\begin{bmatrix} \times & \times & \times & \times \\ & \times & \times & \times \\ & & \times & \times \\ & & & \times \end{bmatrix} \begin{Bmatrix} \delta_1 \\ \delta_2 \\ \delta_4 \\ \delta_3 \end{Bmatrix} = \begin{Bmatrix} F_1 \\ F_2 \\ F_4 \\ F_3 \end{Bmatrix}$$

(i) δ_1 is eliminated,
   5 sets of coefficients written to store,
   14 sets of coeffients modified.
Because the first row and column now contain only zero coefficients the front area contracts.
(ii) δ_2 is eliminated,
   4 sets of coefficients written to store,
   9 sets of coefficients modified.
Second, element 2 is assembled and reduced.

$$\begin{bmatrix} \times & \times & \times & \times \\ & \times & \times & \times \\ & & \times & \times \\ & & & \times \end{bmatrix} \begin{Bmatrix} \delta_5 \\ \delta_6 \\ \delta_4 \\ \delta_3 \end{Bmatrix} = \begin{Bmatrix} F_5 \\ F_6 \\ F_4 \\ F_3 \end{Bmatrix}$$

(i) δ_5 is eliminated,
      5 sets of coefficients written to store,
      14 sets of coefficients modified and front area contracts.
(ii) δ_6 is eliminated,
      4 sets of coefficients written to store,
      9 sets of coefficients modified and front area contracts further.
(iii) δ_4 is eliminated,
      3 sets of coefficients written to store,
      5 sets of coefficients modified and front area contracts even
      further.
(iv) δ_3 is eliminated,
      2 sets of coefficients written to store,
      2 sets of coefficients modified.

For the front solver using the longevity technique only 76 operations
are required and so even on a small problem there is a saving of
approximately 10% on a simple operations count. The backsubstitution
would likewise be more efficient because of the shorter length of the
equations. However, the saving in time during the backsubstitution
stage is very small compared with the time saved during the reduction.

A word of caution is probably appropriate at this stage. To implement
longevity considerations in the front solver is to proceed along the
same path that occurred in the band solver when complete nodal renumber-
ing schemes were introduced for minimizing the bandwidth and it is not
impossible to spend more execution time in setting up the improved
destination vectors than that which can be saved through a reduction in
the number of operations. The authors' experience suggests that little
advantage is gained using a longevity front solver for two-dimensional
problems but significant savings can be made in three-dimensional
analysis particularly where front widths in excess of 200 occur.

References

1. B. M. Irons. A frontal solution program for finite element analysis.
   International Journal for Numerical Methods in Engineering, vol.2,
   no.1, 1970, pp.5-32.
2. A. K. Gupta and B. M. Mohray. A method for computing numerically
   integrated stiffness matrices. *International Journal for Numerical
   Methods in Engineering*, vol.5, no.1, 1972, pp.83-89.
3. M. F. Yeo. A more efficient front solution: allocating assembly
   locations by longevity considerations. *International Journal for
   Numerical Methods in Engineering*, vol.7, no.4, 1973, pp.570-573.

5 Computer Program Listing and Computations Using Isoparametric Elements

5.1 Computer program listing

Most of the subroutines associated with the eight-node isoparametric element program have been listed in Chapters 3 and 4. The remaining subroutines, which will be described in this chapter, are the master driving program, BLOCK DATA, subroutine INTCRD to interpolate for the nodal coordinates of midside nodes on straight edges and subroutine INDAT which reads the data.

5.1.1 THE MASTER DRIVING PROGRAM

The master program or control program is normally a very simple routine whose main function is to call the various subroutines so that the sequence of operations is performed in the correct order. The secondary function of this routine is to initialize the variables, matrices and vectors to either zero or some predetermined non-zero value.

Program ISOP

```
PROGRAM    ISOP(INPUT,OUTPUT,TAPE5=INPUT,TAPE6=OUTPUT,TAPE2,TAPE3,
1                        TAPE4)
```

```
COMMON/MFY1/INT1(4),INT2(4),INT3(4),XX(8),YY(8)
COMMON/MFY2/WTFUN(3),VECTLC(3)
COMMON RS(16),AW(4),S(136),SHP(8),P(8),CW(4),X(8),Y(8)
COMMON DISP(100,2),TRANS(10,4),PRES(100),REACT(100,2)
COMMON NELDES(8),NELDEF(8),LDEF(10,8)
COMMON DX(8),DY(8),U(8),V(8),SIGMA(3),D(5),YM(5),PR(5),WT(5)
COMMON SRS(150),SS(4000),CORD(100,2),MAT( 10),REQ(2000),LREQ(300)
COMMON NDISP(100,3),DISPL(100,2),NW(150),JW(150),NDISPZ(100,3)
COMMON LDEST(100),NTRANS(100,2),NPRES(100),NREACT(100)
COMMON NVABZ,MAXDIS,NIC,LIV,NNODZ,NSNW,NFNW,MAXNDZ,MAXNEL,NEL,NZN
COMMON MZM,MAXFW,MAXREQ,MAXNW,MAXNOD,MAXMAT,MAXPRS,MAXRCT
COMMON YMOD,PRAT,WEIGHT,WX,WY,XL,YL,PIVOT,DETJ,WIRL,GRAVX,GRAVY
COMMON GRAVTY,NPUT,NRULE,LIN,NSTOP,MAXSS,MAXTNS
```

(1)

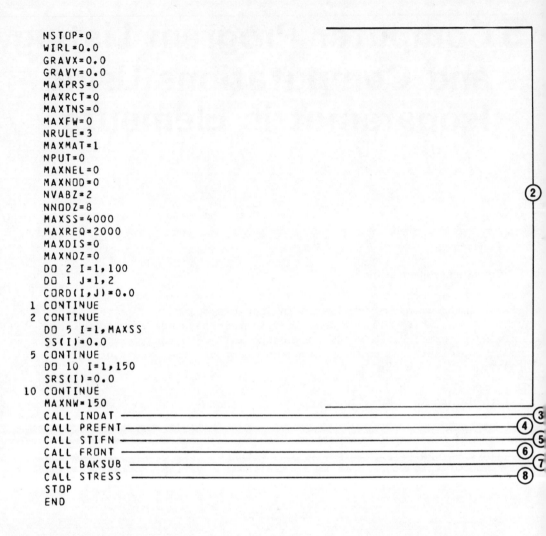

```
        NSTOP=0
        WIRL=0.0
        GRAVX=0.0
        GRAVY=0.0
        MAXPRS=0
        MAXRCT=0
        MAXTNS=0
        MAXFW=0
        NRULE=3
        MAXMAT=1
        NPUT=0
        MAXNEL=0
        MAXNOD=0
        NVABZ=2
        NNODZ=8
        MAXSS=4000
        MAXREQ=2000
        MAXDIS=0
        MAXNOZ=0
        DO 2 I=1,100
        DO 1 J=1,2
        CORD(I,J)=0.0
   1    CONTINUE
   2    CONTINUE
        DO 5 I=1,MAXSS
        SS(I)=0.0
   5    CONTINUE
        DO 10 I=1,150
        SRS(I)=0.0
  10    CONTINUE
        MAXNW=150
        CALL INDAT
        CALL PREFNT
        CALL STIFN
        CALL FRONT
        CALL BAKSUB
        CALL STRESS
        STOP
        END
```

(1) Common block which should be inserted in all subroutines.

(2) Initialization of variables, matrices and vectors.

(3) INDAT reads the data for the problem.

(4) PREFNT determines the nodal destinations and the maximum front width.

(5) STIFN calculates the element stiffnesses and load vectors.

(6) FRONT assembles and reduces the structural stiffness matrix and load vector.

(7) BAKSUB calculates the nodal displacements from the reduced equations set up by FRONT.

(8) STRESS determines the element stresses.

5.1.2 BLOCK DATA

Subroutine BLOCK DATA is used, in this case, to initialize vectors efficiently. To carry out this initialization in the master driving program would have necessitated considerably more FORTRAN statements. There is no call to this subroutine as it is called automatically by the computer executive system before execution commences.

Subroutine BLOCK DATA

```
C      INSERT COMMON BLOCK


       DATA XX/1.0,1.0,1.0,0.0,-1.0,-1.0,-1.0,0.0/
       DATA YY/-1.0,0.0,1.0,1.0,1.0,0.0,-1.0,-1.0/
       DATA INT1/1,3,5,7/
       DATA INT2/2,4,6,8/
       DATA INT3/3,5,7,1/
       DATA WTFUN/0.55555555555,0.33333333889,0.55555555556/
       DATA VECTLC/-0.774596669324,0.0,0.774596669324/
       END
```

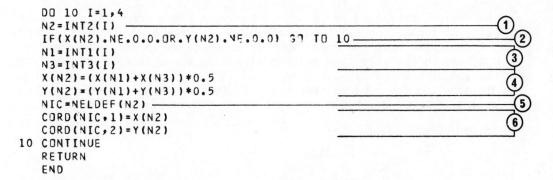

(1) XX( ) and YY( ) contain the local coordinates of each of the eight nodes.
(2) INT2( ) contains the midside node number between the two corner nodes INT1( ) and INT3( ).
(3) WTFUN( ) is the weighting values for the Gauss points, the coordinates of which are defined by VECTLC( ).

5.1.3 NODAL COORDINATE INTERPOLATION SUBROUTINE

Subroutine INTCRD checks that the nodal coordinate of each of the midside nodes in an element have been specified. Coordinates not specified are calculated by a linear interpolation from the coordinates of the adjacent corner nodes. The coordinates are taken to be unspecified if both the x and y coordinates of the node are zero.
   This subroutine is called by subroutine STIFN.

Subroutine INTCRD

```
C      INSERT COMMON BLOCK


       DO 10 I=1,4
       N2=INT2(I)
       IF(X(N2).NE.0.0.OR.Y(N2).NE.0.0) GO TO 10
       N1=INT1(I)
       N3=INT3(I)
       X(N2)=(X(N1)+X(N3))*0.5
       Y(N2)=(Y(N1)+Y(N3))*0.5
       NIC=NELDEF(N2)
       CORD(NIC,1)=X(N2)
       CORD(NIC,2)=Y(N2)
    10 CONTINUE
       RETURN
       END
```

(1) N2 is a midside node.
(2) Checking if the coordinates of N2 are specified.
(3) N1 and N3 are the corner node numbers adjacent to N2.
(4) Interpolating for the coordinates of N2.
(5) NIC is the actual node number of N2.
(6) Storing the coordinates of NIC.

5.1.4 INPUT SUBROUTINE

An input subroutine must be able to serve the following two functions:
(i) to enable the user to prepare the data as easily as possible and
(ii) to present the requisite information to the computer to enable it
to solve the problem. For a finite element program, the data can be
subdivided into five types, each of which has already been discussed
in detail in previous chapters. These five groups of data, which
include element definitions, nodal coordinates, nodal fixity, material
properties and applied loads, are read by subroutine INDAT in the
isoparametric element program. In INDAT, standard format has been
used instead of free format since the latter is machine dependent.
However, apart from this limitation, INDAT has been made quite general
and is capable of accepting the various types and subtypes of data in
any order provided each type or subtype is complete in itself. Inside
each subtype the data may be presented in any order and cards do not
have to be counted beforehand. Furthermore, it is not necessary to
specify the maximum number of elements, highest node number, number of
loaded nodes, etc., since the input subroutine works all these out for
itself.

This method of input is achieved by allocating the first four
columns on every data card to two variables ICODE and NCODE specified
in 2I2 format. Each set of data is then preceded by a card containing
only the first four columns punched specifying values for ICODE and
NCODE. No other information is allowed on this card. The variable
ICODE indicates the type of data to follow, e.g. nodal coordinates,
element definitions, etc., and the variable NCODE indicates the particular subtype of the data, e.g. Cartesian or polar coordinates. Each
card in the subsequent data then has columns 1 to 4 unpunched. To
change the type of data it is only necessary to insert another card
containing a non-zero ICODE. The rest of this card is read by the
program in the same FORMAT as that used to read the current type of
data. However, other than the values of ICODE and NCODE, the rest of
the data on this card is ignored. To avoid field errors in the input
it is therefore necessary to leave blank columns 5 to 80 on this card.

To indicate that the data input is complete it is only necessary to
punch 99 in the first two columns of a card which is then inserted at
the end of all the data.

A detailed input manual for the illustrative program, defining the
values of ICODE and NCODE for each type of data, is given below.

ICODE	NCODE	
01	01	User headings READ ( ) HEAD FORMAT (5X, 9A8) The user may insert as many cards as required with the alphanumeric text to identify the particular analysis.
02	01	Nodal stress averaging READ ( ) NPUT FORMAT (5X, I5) To obtain nodal stress averaging the variable NPUT is set to a non-zero integer value. Default is NPUT = 0 giving no nodal stress averaging.

03   01   Element material properties
            READ (   ) NMAT, YMOD, PRAT, WEIGHT
            FORMAT (5X, I5, 3E10.3)
            NMAT   = material number
            YMOD   = Young's modulus   } for the material
            PRAT   = Poisson's ratio   } number defined
            WEIGHT = Weight per unit volume   } by NMAT

03   02   Material disposition
            READ (   ) NMAT, (NW(I),I=1,14)
            FORMAT (5X, I5, 14I5)
            NMAT   = material number
            NW( )   = the numbers of up to 14 elements made from
                      the material defined by NMAT. The default
                      value for elements omitted from this section
                      is material number 1.

04   01   Cartesian coordinates.
            READ (   ) NIC, COX, COY
            FORMAT (5X, I5, 2E10.3)
            NIC   = node number
            COX   = X Cartesian coordinate
            COY   = Y Cartesian coordinate.

04   02   Polar coordinates
            READ (   ) NIC, RAD, ANG, COX, COY
            FORMAT (5X, I5, 4E10.3)
            NIC   = node number
            RAD   = radius   } for polar coordinates
            ANG   = angle in degrees   }
            COX   = } Cartesian coordinates of origin of polar
            COY   = } coordinate system. *See* Fig.5.1.

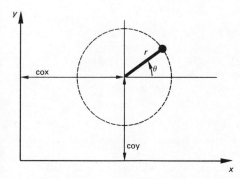

Fig.5.1 Polar coordinates and centre of polar system

05   01   Element definitions
            READ (   ) (NW(I),I=1,8)
            FORMAT (10X, 8I5)
            NW( )   = eight node numbers defining an element. The
                      first element defined in the data list is
                      element number 1. The second element number
                      2, etc.

06   01   Specified nodal displacements.
READ (   ) NFIXX, NFIXY, NIC, DISX, DISY
FORMAT (5X, 2I1, 3X, I5, 5X, 2F10.5)
NFIXX = \ Nodal constraint code
NFIXY = / See Section 2.4.5.
DISX   = specified displacement in x direction
DISY   = specified displacement in y direction
DISX or DISY may be left undefined when the associated
constraint code is zero.

06   02   Earthed nodes.
READ (   ) NFIXX, NFIXY, (NW(I),I=1,14)
FORMAT (5X, 2I1, 3X, 14I5)
NFIXX = \ Nodal constraint code
NFIXY = / See Section 2.4.5.
NW( ) = a list of up to 14 node numbers that have zero
         specified displacements and a constraint code
         defined by NFIXX, NFIXY.

07   01   Nodal point loads
READ (   ) NIC, REX, REY.
FORMAT (5X, I5, 2E10.3)
NIC   = node number
REX   = x component of nodal point load applied at
         node NIC
REY   = y component

08   01   Edge pressure
READ (   ) NIC, PRE
FORMAT (5X, I5, E10.3)
NIC   = node number
PRE   = edge pressure at node NIC

09   01   Body forces
READ (   ) WIRL, GRAVX, GRAVY
FORMAT (5X, 3F10.5)
WIRL   = angular rotation of structure about global
         origins in radians per second
GRAVX = \ components of gravity in the x and y
GRAVY = / directions.

10   01   Transformation matrices
READ (   ) (AW(I),I=1,4)
FORMAT (10X, 4F10.3)
AW( ) = direction cosine matrix (see example problem)
The first transformation matrix in the data is matrix
No.1. The second is No.2, etc. The matrix is presented
in the form

$$|L| = \begin{bmatrix} 1,1 & 1,2 \\ 2,1 & 2,2 \end{bmatrix} \rightarrow AW( )$$

$$= \begin{bmatrix} L(1,1), & L(1,2), & L(2,1), & L(2,2) \end{bmatrix}$$

10	02	Nodes with locally defined axes (*see* Section 4.3)
		READ ( ) NTR, (NW(I),I=1,14)
		FORMAT (5X,I5, 14I5)
		NTR = transformation matrix number
		NW( ) = a list of up to 14 node numbers whose displacements are to be defined relative to the local coordinate system defined by transformation matrix number NTR.
99		Data complete

Subroutine INDAT

```
C    INSERT COMMON BLOCK

     DIMENSION HEAD(9)
   1 READ(5,1002) ICODE,NCODE
   2 WRITE(6,1083) ICODE,NCODE                                        ①
     IF(ICODE.EQ.99) RETURN                                          ②
     GO TO(100,150,200,250,300,350,400,450,500,550),ICODE           ③
 100 WRITE(6,2001)
 101 READ(5,1002) ICODE,NCODE,HEAD
     IF(ICODE.NE.0) GO TO 2
     WRITE(6,1003) HEAD                                              ④
     GO TO 101
 150 WRITE(6,2002)
     READ(5,1004) NPUT
     WRITE(6,1005) NPUT                                              ⑤
     GO TO 1
 200 GO TO(201,210),NCODE
 201 WRITE(6,2003)
 202 READ(5,1006) ICODE,NCODE,NMAT,YMOD,PRAT,WEIGHT
     IF(ICODE.NE.0) GO TO 2
     WRITE(6,1007) NMAT,YMOD,PRAT,WEIGHT
     YM(NMAT)=YMOD                                                   ⑥
     PR(NMAT)=PRAT
     WT(NMAT)=WEIGHT
     GO TO 202
 210 WRITE(6,2004)
 211 READ(5,1014) ICODE,NCODE,NMAT,(NW(I),I=1,14)
     IF(ICODE.NE.0) GO TO 2
     WRITE(6,1015) NMAT,(NW(I),I=1,14)
     IF(MAXMAT.LT.NMAT) MAXMAT=NMAT
     DO 220 I=1,14                                                   ⑦
     NEL=NW(I)
     IF(NEL.EQ.0) GO TO 220
     MAT(NEL)=NMAT
 220 CONTINUE
     GO TO 211
 250 GO TO(251,260),NCODE
 251 WRITE(6,2005)
 252 READ(5,1006) ICODE,NCODE,NIC,COX,COY
     IF(ICODE.NE.0) GO TO 2
     WRITE(6,1007) NIC,COX,COY                                       ⑧
     CORD(NIC,1)=COX
     CORD(NIC,2)=COY
     GO TO 252
 260 WRITE(6,2006)
 261 READ(5,1006) ICODE,NCODE,NIC,RAD,ANG,COX,COY
     IF(ICODE.NE.0) GO TO 2
     WRITE(6,1007) NIC,RAD,ANG,COX,COY                               ⑨
     A1=0.0174532925*ANG
     CORD(NIC,1)=RAD*COS(A1)+COX
     CORD(NIC,2)=RAD*SIN(A1)+COY
     GO TO 261
```

```
300 NEL=0
    WRITE(6,2007)
301 NEL=NEL+1
    READ(5,1018) ICODE,NCODE,(NW(I),I=1,8)
    IF(ICODE.NE.0) GO TO 303
    DO 302 I=1,8
    LDEF(NEL,I)=NW(I)
302 CONTINUE
    WRITE(6,1015) NEL,(LDEF(NEL,I),I=1,NNODZ)
    MAXNEL=NEL
    GO TO 301
303 DO 305 NEL=1,MAXNEL
    IF(MAT(NEL).EQ.0) MAT(NEL)=1
    DO 304 LNOD=1,NNODZ
    IF(MAXNOD.LT.LDEF(NEL,LNOD)) MAXNOD=LDEF(NEL,LNOD)
304 CONTINUE
305 CONTINUE
    GO TO 2
350 NCOUNT=0
    GO TO(351,360),NCODE
351 WRITE(6,2008)
352 NCOUNT=NCOUNT+1
    READ(5,1026) ICODE,NCODE,NFIXX,NFIXY,NIC,DIS1,DIS2
    IF(ICODE.NE.0) GO TO 2
    NDISP(NCOUNT,1)=NFIXX
    NDISP(NCOUNT,2)=NFIXY
    NDISP(NCOUNT,3)=NIC
    DISP(NCOUNT,1)=DIS1
    DISP(NCOUNT,2)=DIS2
    WRITE(6,1027) (NDISP(NCOUNT,I),I=1,3),(DISP(NCOUNT,I),I=1,2)
    MAXDIS=NCOUNT
    GO TO 352
360 WRITE(6,2009)
361 READ(5,1008) ICODE,NCODE,NFIXX,NFIXY,(NW(I),I=1,14)
    IF(ICODE.NE.0) GO TO 2
    WRITE(6,1009) NFIXX,NFIXY,(NW(I),I=1,14)
    DO 365 I=1,14
    IF(NW(I).EQ.0) GO TO 365
    NCOUNT=NCOUNT+1
    NDISPZ(NCOUNT,3)=NW(I)
    NDISPZ(NCOUNT,1)=NFIXX
    NDISPZ(NCOUNT,2)=NFIXY
    MAXNDZ=NCOUNT
365 CONTINUE
    GO TO 361
400 NCOUNT=0
    WRITE(6,2010)
401 NCOUNT=NCOUNT+1
    READ(5,1006) ICODE,NCODE,NIC,REX,REY
    IF(ICODE.NE.0) GO TO 2
    NREACT(NCOUNT)=NIC
    REACT(NCOUNT,1)=REX
    REACT(NCOUNT,2)=REY
    WRITE(6,1007) NREACT(NCOUNT),(REACT(NCOUNT,I),I=1,2)
    MAXRCT=NCOUNT
    GO TO 401
450 NCOUNT=0
    WRITE(6,2011)
451 NCOUNT=NCOUNT+1
    READ(5,1006) ICODE,NCODE,NIC,PRE
    IF(ICODE.NE.0) GO TO 2
    NPRES(NCOUNT)=NIC
    PRES(NCOUNT)=PRE
    WRITE(6,1007) NPRES(NCOUNT),PRES(NCOUNT)
    MAXPRS=NCOUNT
    GO TO 451
```

(10)

(11)

(12)

(13)

(14)

(15)

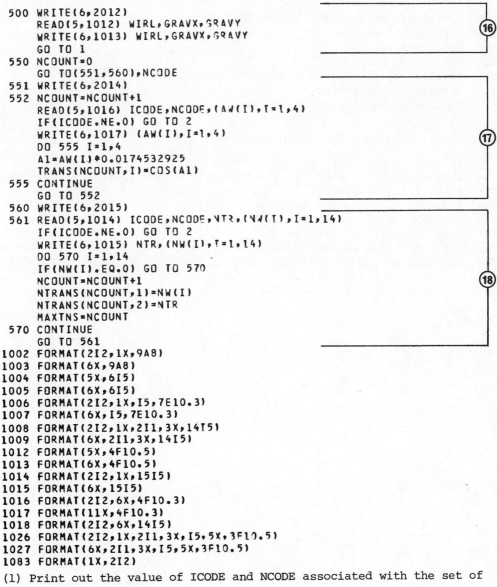

```
 500 WRITE(6,2012)
     READ(5,1012) WIRL,GRAVX,GRAVY
     WRITE(6,1013) WIRL,GRAVX,GRAVY
     GO TO 1
 550 NCOUNT=0
     GO TO(551,560),NCODE
 551 WRITE(6,2014)
 552 NCOUNT=NCOUNT+1
     READ(5,1016) ICODE,NCODE,(AW(I),I=1,4)
     IF(ICODE.NE.0) GO TO 2
     WRITE(6,1017) (AW(I),I=1,4)
     DO 555 I=1,4
     A1=AW(I)*0.0174532925
     TRANS(NCOUNT,I)=COS(A1)
 555 CONTINUE
     GO TO 552
 560 WRITE(6,2015)
 561 READ(5,1014) ICODE,NCODE,NTR,(NW(I),I=1,14)
     IF(ICODE.NE.0) GO TO 2
     WRITE(6,1015) NTR,(NW(I),I=1,14)
     DO 570 I=1,14
     IF(NW(I).EQ.0) GO TO 570
     NCOUNT=NCOUNT+1
     NTRANS(NCOUNT,1)=NW(I)
     NTRANS(NCOUNT,2)=NTR
     MAXTNS=NCOUNT
 570 CONTINUE
     GO TO 561
1002 FORMAT(2I2,1X,9A8)
1003 FORMAT(6X,9A8)
1004 FORMAT(5X,6I5)
1005 FORMAT(6X,6I5)
1006 FORMAT(2I2,1X,I5,7E10.3)
1007 FORMAT(6X,I5,7E10.3)
1008 FORMAT(2I2,1X,2I1,3X,14I5)
1009 FORMAT(6X,2I1,3X,14I5)
1012 FORMAT(5X,4F10.5)
1013 FORMAT(6X,4F10.5)
1014 FORMAT(2I2,1X,15I5)
1015 FORMAT(6X,15I5)
1016 FORMAT(2I2,6X,4F10.3)
1017 FORMAT(11X,4F10.3)
1018 FORMAT(2I2,6X,14I5)
1026 FORMAT(2I2,1X,2I1,3X,I5,5X,3F10.5)
1027 FORMAT(6X,2I1,3X,I5,5X,3F10.5)
1083 FORMAT(1X,2I2)
```

(1) Print out the value of ICODE and NCODE associated with the set of data to follow.
(2) If ICODE = 99 then all the data has been read.
(3) The value of ICODE determines the type of data to follow and hence control is transferred to that part of the subroutine capable of reading that data.
(4) Section to read user alphanumeric headings.
(5) NPUT is the flag to indicate request for average nodal stresses.
(6) Material properties.
(7) Material disposition.
(8) Nodal Cartesian coordinates.
(9) Nodal polar coordinates read in and transformed to Cartesian coordinates before storage.
(10) Element definitions and determination of number of elements.
(11) Determination of highest node number and setting default value for undefined element material disposition.

(12) Specified non-zero displacements.
(13) Earthed nodes.
(14) Nodal point loads.
(15) Nodal edge pressures.
(16) Angular velocity and components of gravity.
(17) Transformation matrices.
(18) Nodes with locally defined axes.

5.2 Example problem using isoparametric elements

To enable a comparison to be made between the elementary program
described in Chapter 2 and the isoparametric element program, the same
problem of a thick cylinder subjected to internal pressure will be
analysed. Also to provide a direct comparison between the two programs
the same quarter of a cylinder will be analysed using the two mesh
divisions shown in Fig.5.2(a), (b). To illustrate the use of the
facility in the isoparametric program allowing nodes to be given locally
defined axes, two additional mesh divisions will also be analysed (Fig.
5.2(c), (d)) which represent a 6° segment of the cylinder. The 6°
segment was arbitrarily chosen.

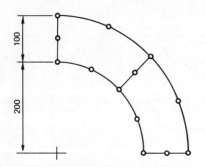

(a)   Course mesh for quarter cylinder

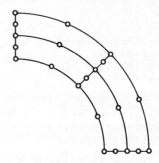

(b)   Fine mesh for quarter cylinder

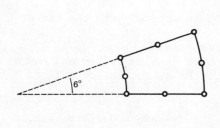

(c)   Course mesh for 6° segment

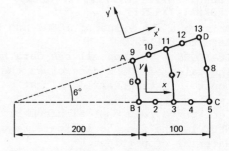

(d)   Fine mesh for 6° segment

Fig.5.2 Isoparametric mesh divisions for a thick cylinder

5.3 Input data for the 6° segment

To minimize the possibility of errors in data generation the first task
is to draw the structure to scale and then subdivide it into elements.
Once this has been done it is possible to locate the nodes and to give
each a number as shown in Fig.5.2(d).

The nodal coordinates can be specified in either Cartesian or polar
coordinates. For this mesh it is obviously easier to use polar coordinates with the origin at the axis of tne cylinder. For example, for node
9 the data card would contain the following information

NODE	RAD	ANG	COX	COY
9	200.0	6	0.0	0.0

COX and COY are the Cartesian coordinates of the centre of the polar
system (see Fig.5.1).

If any of the midside nodal coordinates are omitted from the data list
then subroutine INTCRD automatically carries out a linear interpolation
for these coordinates from the coordinates of the adjacent corner nodes.
Hence the coordinates of nodes 2, 4, 10 and 12 which are situated on
straight edges can be omitted from the input, and only midside nodes
situated on curved edges need to have their coordinates specified.

The element definitions for the two elements are:

Element 1	3	4	5	8	13	12	11	7
Element 2	1	2	3	7	11	10	9	6

The material properties associated with these elements are:

Young's modulus 0.1×10^6 MPa

Poisson's ratio   0.25

Since there is only one material this can be material No.1. This is
the default number for element material properties and therefore the
material disposition does not have to be specified.

The material properties are input as

NMAT	YMOD	PRAT	WEIGHT
1	0.100E06	0.250	0.0

The problem does not involve body forces so the weight per unit volume
can be omitted and is zero by default.

All specified displacements for this problem are zero displacements.
However, before all the fixities can be introduced the displacements of
all the nodes along AD in Fig.5.2(d) must be locally orientated, so that
the axisymmetric situation can be modelled by having roller supports
along AD. To enforce $V = 0$ for nodes on BC and $V' = 0$ for nodes on AD
a constraint code of 01 is required. This data can be input on one
card as

Constraint code	Nodes associated with the constraint code
01	1 2 3 4 5 9 10 11 12 13

The only applied loading is an edge pressure of 100 MPa on face AB, i.e. on nodes 1, 6 and 9. This is input as

    1   0.100 E03

    6   0.100 E03

    9   0.100 E03

It only remains to define the local coordinate system for AD to complete the data. For this purpose, it is desirable to draw the two coordinate systems superimposed and the definition of the DC matrix, i.e.

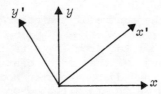

$$[L] = \begin{bmatrix} XX' & XY' \\ YX' & YY' \end{bmatrix}$$

The angles corresponding to YX' for example are measured in an anti-clockwise direction from the first axis Y to the second axis X', i.e.

$$[L] = \begin{bmatrix} \cos 6^\circ & \cos 96^\circ \\ \cos 276^\circ & \cos 6^\circ \end{bmatrix} = \begin{bmatrix} \cos 6^\circ & -\sin 6^\circ \\ \sin 6^\circ & \cos 6^\circ \end{bmatrix}$$

This DC matrix is input as

XX' XY' YX' YY'

6.0   96.0   276.0   6.0

The nodes affected by this transformation matrix are

9   10   11   12   13

Since only one DC matrix has been defined in the input it is DC matrix no.1. The transformation data is therefore input as

DC matrix no.   Nodes associated with this DC matrix

        1   9   10   11   12   13

The end of data is indicated by punching 99 in the first two columns of a card and inserting at the end of the data.

Program data echo

```
1 1      USER HEADINGS
    TWO ELEMENT 6-DEG WEDGE
3 1       MATERIAL PROPERTIES
       1 .100E+06  .250E+00   .100E+03
```

```
4 2       POLAR CO-ORDINATES
          1   .200E+03 0.          0.          0.
          6   .200E+03  .300E+01 0.          0.
          9   .200E+03  .600E+01 0.          0.
          3   .250E+03 0.          0.          0.
          7   .250E+03  .300E+01 0.          0.
         11   .250E+03  .600E+01 0.          0.
          5   .300E+03 0.          0.          0.
          8   .300E+03  .300E+01 0.          0.
         13   .300E+03  .600E+01 0.          0.
5 1       ELEMENT DEFINITIONS
          1   3    4    5    8   13   12   11    7
          2   1    2    3    7   11   10    9    6
6 2       EARTHED NODES
     01   1    2    3    4    5    9   10   11   12   13    0    0    0    0
8 1       EDGE PRESSURES
          1  .100E+03
          6  .100E+03
          9  .100E+03
10 1      TRANSFORMATION MATRICES
             6.000    96.000   276.000      6.000
10 2      TRANSFORMED NODES
          1   9   10   11   12   13    0    0    0    0    0    0    0    0
99 0
```

Once the input has been accepted the first operation carried out automatically by the program is to determine the nodal destination vector and the front width. The front width for this problem is 16.

The information required by the front solver is then set up by subroutine STIFN, i.e. the element stiffness, load vectors, definition vector and destination vectors. The element stiffness is stored in the S( ) vector. For element 1 the S( ) vector written in the form

        WRITE (6,5001) S

  5001 FORMAT (40 (/, 5(2X, E10.3)))

contains the following coefficients

```
.755E+05    .297E+05    .127E+06   -.476E+05   -.192E+05
-.103E+05   -.718E+04    .323E+05    .351E+03   -.187E+04
.421E+05   -.336E+05   -.470E+04   -.470E+04   -.836E+05
.333E+05    .843E+04    .175E+05    .530E+05   -.330E+05
-.584E+04   -.117E+05   -.228E+05    .346E+05   -.624E+04
.624E+04    .628E+05   -.617E+05   -.619E+04   -.151E+05
-.172E+06    .146E+06   -.144E+04    .140E+06   -.525E+05
.104E+05    .193E+05   -.882E+04   -.653E+03   -.323E+05
-.323E+05   -.276E+04   -.306E+05   -.331E+04   -.886E+04
-.221E+05    .176E+05    .318E+04   -.318E+04   -.781E+05
-.330E+05    .117E+05    .584E+04   -.228E+05    .122E+04
.269E+05    .269E+05    .220E+04    .763E+05   -.330E+05
.114E+06   -.546E+05    .143E+05    .231E+05   -.155E+06
.315E+05   -.373E+04    .373E+04    .607E+05   -.306E+05
.886E+04    .331E+04   -.221E+05    .333E+05   -.175E+05
-.843E+04    .530E+05   -.369E+05    .101E+05    .101E+05
-.838E+05    .154E+06   -.900E+04    .325E+06   -.528E+05
-.684E+04   -.199E+05   -.156E+06   -.402E+04    .324E+05
.320E+05    .627E+03   -.339E+05    .134E+05    .116E+04
-.827E+05    .296E+05   -.644E+04   -.644E+04    .152E+06
.763E+05    .330E+05    .114E+06   -.525E+05   -.193E+05
-.104E+05   -.882E+04    .323E+05    .187E+04   -.351E+03
.421E+05   -.356E+05   -.139E+05   -.125E+04   -.844E+05
.146E+06    .144E+04    .140E+06   -.476E+05    .103E+05
.192E+05   -.718E+04    .402E+04   -.266E+05   -.270E+05
-.627E+03    .755E+05   -.297E+05    .127E+06   -.620E+05
-.287E+03    .330E+05   -.169E+06    .161E+06   -.104E+05
.358E+06
```

This represents the upper triangle of the element stiffness shown in Fig.5.3.

```
75500.  29700. -47600. -19200.  32300.    351. -33600.  -4700.  33300.   8430. -33000.  -5840.  34600.  -6240. -61700.  -6190.
       127000. -10300.  -7180.  -1870.  42100.  -4700. -83600.  17500.  53000. -11700. -22800.   6240.  62800. -15100.-172000.
               146000.  -1440. -52500.  10400.   -653. -32300. -30600.  -3310.  17600.   3180. -33000.  11700.   1220.  26900.
                       140000.  19300.  -8820. -32300.  -2760.  -8860. -22100.  -3180.   5840. -22800.  26900.   2200.
                               76300. -33000. -54600.  14300.  31500.  -3730. -30600.   8860. -17500. -36900.  10100.
                                      114000.  23100.-155000.   3730.  60700.   3310. -22100.  53000.  10100. -83800.
                                              154000.  -9000. -52800.  -6840.  -4020.  32400.  13400.  29600.  -6440.
                                                      325000. -19900.-156000.  32000. -33900. -82700.  -6440. 152000.
                                                              76300.  33000. -52500.    627.   1160. -35600. -13900.
                                                                     114000. -10400. -19300.  32300.   1870. -1250. -84400.
                                                                             146000.  1440.  -8820.   -351.  42100.  4020. -26600.
                                                                                     140000. -47600.  10300.  -7180.  10300.  -627.
                                                                                             140000.  19200.  -7180. -27000.  -287.
                                                                                                     75500. -29700. -62000.-169000.
                                                                                                            127000.  33000. -10400.
                                                                                                                    161000.358000.
```

Fig.5.3 Upper triangle of the stiffness matrix of element No.1

```
-.476E+05  -.103E+05  .146E+06  -.144E+04  -.575E+05  .104E+05  -.653E+03  -.323E+05  -.306E+05  -.331E+04
 .176E+05   .318E+04  -.330E+05  .117E+05   .122E+04   .269E+05  0.
 NSNW =  1   NFNW =  16  PIVOT POSITION =   3

-.197E+05  -.728E+04  0.         .100E+51   .188E+05  -.871E+04  -.323E+05  -.308E+04  -.916E+04  -.222E+05
-.301E+04  -.781E+05   .552E+04  -.227E+05   .263E+05   .246E+04  0.
 NSNW =  1   NFNW =  16  PIVOT POSITION =   4

-.153E+05  -.556E+04  0.         .215E+05   .575E+05  -.293E+05  -.548E+05   .270E+04   .205E+05  -.492E+04
-.243E+05   .100E+05  -.133E+05  -.354E+05   .197E+05  0.
 NSNW =  1   NFNW =  16  PIVOT POSITION =   5

 .115E+05   .400E+05  0.         .482E+04   0.        -.100E+51  -.471E+04  -.151E+06   .164E+05   .584E+05
-.103E+05  -.173E+05   .454E+05  -.955E+04  -.757E+05  0.
 NSNW =  1   NFNW =  16  PIVOT POSITION =   6

-.193E+05  -.100E+05  0.        -.136E+05   0.         0.         .101E+06  -.657E+04  -.334E+05  -.115E+05
-.271E+05   .419E+05   .784E+03  -.515E+04   .125E+05  0.
 NSNW =  1   NFNW =  16  PIVOT POSITION =   7

-.171E+05  -.863E+05  0.        -.799E+04   0.         0.         0.         .317E+06  -.298E+05  -.158E+06
 .353E+05  -.353E+04  -.794E+05  -.473E+04   .158E+06  0.
 NSNW =  1   NFNW =  16  PIVOT POSITION =   8

-.993E+04   .593E+04  0.         .125E+05   0.         .363E+04   0.         0.         .488E+05   .155E+05
-.457E+05  -.805E+04   .188E+04  -.245E+05   0.
 NSNW =  1   NFNW =  16  PIVOT POSITION =   9

-.520E+04   .641E+04  0.        -.876E+04   0.        -.341E+04   0.         0.         0.         .100E+51
-.169E+05   .121E+04   .124E+04   .455E+03  0.
 NSNW =  1   NFNW =  16  PIVOT POSITION =  10

-.148E+05  -.351E+03  0.        -.256E+05   0.        -.324E+05   0.         0.         0.
 .801E+05   .855E+04   .141E+05  -.355E+05  0.
 NSNW =  1   NFNW =  16  PIVOT POSITION =  11

 .391E+04  -.155E+05  0.         .267E+05   0.        -.752E+04   0.         0.         0.
 0.         .100E+51  -.574E+04  -.189E+05  0.
 NSNW =  1   NFNW =  16  PIVOT POSITION =  12
```

Fig.5.4 Stored equations after assembly of element No.1

```
-.318E+05   .280E+02   .791E+05  -.364E+05  -.188E+05  -.293E+05  -.627E+04   .377E+05  -.709E+04
-.778E+05  -.522E+04   .344E+05  -.399E+05  -.411E+04   .349E+03
NSNW =  1   NFNW =  16   PIVOT POSITION =  3

-.139E+05   .486E+05   .100E+51   .347E+04   .686E+04  -.113E+04  -.220E+05   .678E+04   .785E+05
 .145E+05  -.211E+06   .596E+05   .105E+05  -.100E+06  -.129E+03
NSNW =  1   NFNW =  16   PIVOT POSITION =  4

-.271E+05   .108E+05  0.        -.118E+06  -.112E+05  -.140E+05   .259E+04  -.119E+05   .865E+04
-.345E+05  -.239E+05  -.107E+05   .415E+03  -.199E+05  -.347E+05   .161E+03
NSNW =  1   NFNW =  16   PIVOT POSITION =  5

 .247E+05  -.180E+04  0.        0.         .100E+51  -.138E+05  -.105E+06   .141E+05  -.252E+05
 .451E+04   .314E+04  -.120E+04  -.206E+05  -.441E+05  -.708E+04   .984E+02
NSNW =  1   NFNW =  16   PIVOT POSITION =  6

-.106E+06  -.111E+05  0.        0.        0.        -.179E+05   .115E+05   .165E+05  -.132E+05
-.197E+05   .183E+05   .255E+05  -.834E+04  -.115E+06   .103E+05  -.104E+03
NSNW =  1   NFNW =  16   PIVOT POSITION =  1

 .100E+51  0.        0.         .262E+04   .308E+06  -.216E+05  -.508E+04   .603E+05   .118E+05
-.102E+06   .890E+04   .111E+06  -.340E+04  -.257E+02
NSNW =  2   NFNW =  16   PIVOT POSITION =  2

-.398E+05  -.422E+05  -.568E+04  -.133E+04  -.159E+05   .129E+05  -.775E+05   .215E+05   .144E+06  -.157E+05
 .931E+02
NSNW =  7   NFNW =  16   PIVOT POSITION =  15
```

```
.244E+05   .456E+04  -.385E+04  -.974E+05  -.215E+05   .197E+06  -.416E+05  -.308E+06   0.
.858E+02
NSNW =  7   NFNW =  16   PIVOT POSITION =  16

-.458E+05   .361E+04   .704E+04   .209E+04  -.163E+05   .489E+04   .579E+05   .341E+04  -.565E+02
NSNW =  7   NFNW =  14   PIVOT POSITION =  13

.114E+05  -.561E+04  -.389E+04   .602E+03  -.807E+03  -.619E+04   0.        .100E+51  -.122E+02
NSNW =  7   NFNW =  14   PIVOT POSITION =  14

.713E+05   .168E+05  -.169E+05   .111E+05  -.486E+05  -.222E+05   .109E+03
NSNW =  7   NFNW =  12   PIVOT POSITION =  7

.100E+51   .255E+05  -.103E+04  -.123E+05  -.384E+04  -.146E+02
NSNW =  8   NFNW =  12   PIVOT POSITION =  8

.522E+05  -.210E+05  -.516E+05  -.506E+04   .252E+03
NSNW =  9   NFNW =  12   PIVOT POSITION =  9

.100E+51   .133E+05  -.178E+06   .107E+03
NSNW =  10   NFNW =  12   PIVOT POSITION =  10

.464E+04  -.186E+05   .209E+04
NSNW =  11   NFNW =  12   PIVOT POSITION =  11

.284E+06   .847E+04
NSNW =  12   NFNW =  12   PIVOT POSITION =  12

.629E+06
```

Fig.5.5 Stored equations after assembly of element No.2

After element 1 has been assembled the variables relating to nodes
that do not appear in element 2 can be used for reduction. The
coefficients of the reduced equations, including the control parameters,
stored by the front solver are shown in Fig.5.4. The last coefficient
in each equation is the load term. After assembly of element 2 the
reduced equations stored are shown in Fig.5.5.

The effect of the heading vector on the equation length can be seen
from the reduced equations after assembly of the second element.

The constant length of the reduced equations, before assembly of the
second element, would not have occurred if a longevity prefront had been
implemented. The nodes causing the non-zero coefficients at the end of
the equation would then have been allocated storage towards the left-
hand side of the stiffness matrix which would enable the equations to
contract in length.

The stored reduced equations are then used for backsubstitution to
calculate the displacements. During this operation the reactions at
fixed nodes are calculated. However as the backsubstitution proceeds
the displacements and forces at transformed nodes are in the locally
orientated directions and hence *the reactions at these nodes will also
be in a locally orientated direction.*

Reactions printed out by the program are

REACTIONS ELEMENT   2

NODE	1ST COMP	2ND COMP
9	0.	.215E+04
10	0.	.744E+04
11	0.	.323E+04
3	0.	-.323E+04
2	0.	-.744E+04
1	0.	-.215E+04

REACTIONS ELEMENT   1

NODE	1ST COMP	2ND COMP
12	0.	.585E+04
13	0.	.133E+04
5	0.	-.133E+04
4	0.	-.585E+04

After completion of the backsubstitution the locally orientated
displacements are reorientated into the global direction and the global
displacements of all nodes are printed out.

NODAL DISPLACEMENTS

NODE	X-COMP	Y-COMP
1	.5699737	0.0000000
2	.5349848	0.0000000
3	.5099799	0.0000000
4	.4922556	0.0000000
5	.4799816	0.0000000
6	.5691878	.0298299
7	.5092776	.0266901
8	.4793219	.0251202
9	.5668513	.0595785
10	.5320541	.0559211
11	.5071862	.0533074
12	.4895590	.0514547
13	.4773522	.0501717

The final operation in the program is to use the global displacements to calculate the element stresses which are printed out.

STRESSES ELEMENT   1

NODE	SIGMA X-X	SIGMA Y-Y	TAU X-Y
3	-32.9	195.8	-.0
4	-16.3	174.9	.0
5	1.9	160.5	-.0
8	2.3	160.0	-8.3
13	3.7	158.7	-16.5
12	-14.2	172.8	-19.9
11	-30.4	193.3	-23.7
7	-32.2	195.2	-11.9

STRESSES ELEMENT   2

NODE	SIGMA X-X	SIGMA Y-Y	TAU X-Y
1	-94.6	261.3	-.0
2	-64.6	221.6	.0
3	-31.0	196.2	-.0
7	-30.4	195.6	-11.9
11	-28.5	193.8	-23.6
10	-61.5	218.5	-29.8
9	-90.7	257.5	-37.0
6	-93.6	260.4	-18.6

The results obtained from the four meshes used to represent the thick cylinder subjected to 100 MPa internal pressure are given in Table 5.1. It can be seen that each of the meshes give good agreement with the theoretical tangential stress. However there must be at least two elements in the radial direction, meshes (b) and (d), before reasonable agreement is obtained with the theoretical radial stress. Nevertheless comparing the results obtained using the constant strain triangle with those obtained using the isoparametric element illustrates the efficiency of the second element compared with the first.

TABLE 5.1 *Comparison of the stresses obtained for a thick cylinder under internal pressure using different meshes*

Radius	Theoretical stress	Mesh (Fig.5.2)			
		(a)	(b)	(c)	(d)
Radial stress (MPa)					
200	-100.0	-86.7	-96.6	-83.4	-96.4
225	- 62.2		-64.4		-64.5
250	- 35.2	-41.4	-31.3	-41.4	-31.8
275	- 15.2		-16.1		-16.2
300	0	13.4	3.1	10.2	1.9
Tangential stress (MPa)					
200	260.0	262.8	260.9	263.9	261.3
225	222.2		222.0		221.6
250	195.2	193.9	196.2	193.5	195.9
275	175.2		175.1		174.9
300	160.0	163.1	160.8	162.4	160.4

6 Automatic Mesh Generation

6.1 Introduction

The example problem used in this book is a very simple one and little difficulty has been encountered in the data preparation. However, for many practical problems, thousands of elements and nodes are involved and the task of preparing the data becomes extremely lengthy and tedious. Moreover, during the preparation of thousands of data cards some human error may be introduced and remain undetected in spite of the checks which are usually made. The presence of such errors will inevitably bring about incorrect results and if detected at that stage it would mean another run on the computer after correcting the data, which is tiresome, but not disastrous. However, if the errors stay undetected and such incorrect results are used for making decisions and judgements, there may be serious repercussions. It is therefore important to eliminate such data error, and this can be achieved to a large extent by automatic mesh generation, in which nodal numbers and their coordinates, together with element numbers and their definitions, are prepared automatically by the computer, using as input data the minimal amount of information necessary to describe the geometry of the domain and the desired fineness of the mesh divisions.

For non-linear problems, which sometimes involve the creation of new mesh divisions based on the results of the previous analysis, automatic mesh generation is essential, especially for problems involving many steps or iterations.

In the last few years, considerable effort has been expended in developing mesh generation routines in order to eliminate the drudgery of working out the data and at times punching out the data cards, and to minimize data errors. An early survey of a number of routines was given by Buell and Bush [1].

6.2 Mesh generation

6.2.1 TRIANGULAR ELEMENTS

An elementary routine for generating a triangular mesh is presented here. The input data consists of information concerning the total number of generating lines (NY), a weighting factor (CON), and for each generating

line the number of intervals (NX(I)) required, together with the coordinates of the two end points (XF(I), YF(I), XL(I), YL(I)). The output
consists of firstly the x and y coordinates of each point with its
corresponding number, and secondly the element number with its element
definition. Both nodal numbers and element numbers are arranged in
sequential order.

To optimize the band width the generating lines should always traverse
the shorter direction of the domain.

The number of divisions in adjacent generating lines can be equal or
differ by one, so that the mesh can vary according to the specific
requirement of the problem. On Fig.6.1, there are two divisions in line
A, but the neighbouring line B is allowed to have one, two or three
divisions.

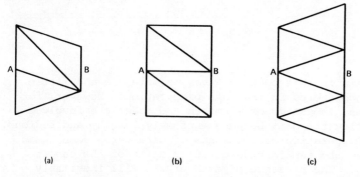

      (a)   (b)   (c)

Fig.6.1 Possible triangular element divisions

Another refinement incorporated into the program is the weighting
factor, CON. Depending on whether the weighting factor is < 1, = 1 or
> 1, the intervals along a generating line will become progressively
shorter, stay equal, or become progressively longer.

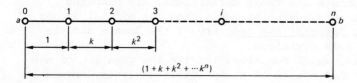

Fig.6.2 Effect of weighting factor upon nodal intervals

From Fig.6.2 it can be seen that the coordinates of a point i along
a generating line ab can be computed by

$$x_i = (x_b - x_a)\frac{\displaystyle\sum_{j=1}^{i} k^{j-1}}{\displaystyle\sum_{j=1}^{n} k^{j-1}} + x_a$$

or, summing the geometric progression,

$$x_i = (x_b - x_a)\frac{k^i - 1}{k^n - 1} + x_a$$

and

$$y_i = (y_b - y_a)\frac{\sum\limits_{j=1}^{i} k^{j-1}}{\sum\limits_{j=1}^{n} k^{j-1}} + y_a = (y_b - y_a)\frac{k^i - 1}{k^n - 1} + y_a$$

The node numbers will be labelled correctly as we go from one generating line to the next, by always increasing the number by one every time we come across a point.

The element definitions are worked out by taking the mth and the $(m + 1)$th generating lines, starting from $m = 1$, and forming n quadrilaterals along the jth line. Referring to Fig.6.1(a), in which the $(m + 1)$th line has one less division than the mth line, the last (nth) quadrilateral degenerates into a triangle, while, for Fig.6.1(b), the mth and the $(m + 1)$th lines have the same number of intervals and each of the n quadrilaterals is simply split into two triangles. For Fig. 6.1(c), however, the $(m + 1)$th line has now one more division than the jth line, and one extra triangle must be added to the n quadrilaterals which have been established.

A number of features can be added to the subroutine without a great deal of difficulty. The shape of the triangles can be improved by always using the shorter diagonal of a quadrilateral when splitting it into two triangles. For layered materials with intermediate boundaries it is only necessary to incorporate intermediate points into a generating line at the interfaces so as to ensure that no element will cross over a boundary. For such a case a generating line may take up the form of a polygon with the nodes at the two ends and at the interfaces.

6.2.2 COMPUTER SUBROUTINE

The complete subroutine listing together with the output for a sample problem (Fig.6.3) are given below.

```
      PROGRAM TGEN(INPUT,OUTPUT,TAPE5=INPUT,TAPE6=OUTPUT)
      DIMENSION NX(20),YL(20),XF(20),YF(20),NOD(4),SUM1(15),XL(20)
      READ(5,1000) NY,CON
      WRITE(6,1001) NY,CON
      DO 100 I=1,NY
      READ(5,1000) NX(I),XF(I),YF(I),XL(I),YL(I)
      WRITE(6,1000) NX(I),XF(I),YF(I),XL(I),YL(I)
  100 CONTINUE
      WRITE(6,1431)
      N=0
```

```
      DO 350 I=1,NY
      NXI=NX(I)+1
      SUM1(1)=0.0
      SUM1(2)=1.0
      SUM=1.0
      IF(NXI-2) 190,291,190
  190 DO 250 K=3,NXI
      SUM1(K)=SUM1(K-1)*CON
      SUM=SUM+SUM1(K)
  250 CONTINUE
  291 CONTINUE
      X=XF(I)
      Y=YF(I)
      DO 300 J=1,NXI
      N=N+1
      X=(XL(I)-XF(I))*SUM1(J)/SUM+X
      Y=(YL(I)-YF(I))*SUM1(J)/SUM+Y
      WRITE(6,1430) X,Y,N
  300 CONTINUE
  350 CONTINUE
      N=0
      NSUM=0
      NYI=NY-1
      WRITE(6,1432)
      DO 600 I=1,NYI
      NXI=NX(I)
      DO 500 J=1,NXI
      IF(J-NXI) 379,371,379
  371 IF(NX(I+1)-NX(I)) 380,379,401
  379 NOD(1)=J+NSUM
      NOD(2)=NOD(1)+1
      NOD(3)=NOD(2)+NXI+1
      NOD(4)=NOD(3)-1
      GO TO 412
  380 NOD(1)=NOD(2)
      NOD(2)=NOD(1)+1
      NOD(4)=0
      GO TO 412
  401 NOD(1)=J+NSUM
      NOD(2)=NOD(1)+1
      NOD(3)=NOD(2)+NXI+1
      NOD(4)=NOD(3)-1
      N=N+1
      WRITE(6,1470) N,NOD(1),NOD(2),NOD(3)
      N=N+1
      WRITE(6,1470) N,NOD(1),NOD(3),NOD(4)
      NOD(1)=NOD(2)
      NOD(2)=NOD(3)+1
      NOD(4)=0
  412 N=N+1
      WRITE(6,1470) N,NOD(1),NOD(2),NOD(3)
      IF(NOD(4))433,434,433
  433 N=N+1
      WRITE(6,1470) N,NOD(1),NOD(3),NOD(4)
  434 CONTINUE
  500 CONTINUE
      NSUM=NSUM+NXI+1
  600 CONTINUE
      STOP
 1000 FORMAT(I5,5X,6F10.5)
 1001 FORMAT(*1 CONTROL DATA*,/,I5,5X,6F10.5)
 1431 FORMAT(*1  X CO-ORDINATE   Y CO-ORDINATE   NODAL POINT*)
 1430 FORMAT(2F16.8,I4)
 1432 FORMAT(*1  ELEMENT NO   NODAL POINTS OF THE ELEMENT*)
 1470 FORMAT(I8,5X,3I8)
      END
```

③ ④ ⑤ ⑥ ⑦

CONTROL DATA

5	1.20000			
5	0.00000	4.00000	0.00000	12.00000
5	1.00000	4.00000	2.30000	12.00000
5	2.00000	4.00000	5.00000	12.00000
6	3.00000	4.00000	8.00000	12.00000
7	4.00000	4.00000	12.00000	12.00000

X CO-ORDINATE	Y CO-ORDINATE	NODAL POINT
0.00000000	4.00000000	1
0.00000000	5.07503763	2
0.00000000	6.36508278	3
0.00000000	7.91313696	4
0.00000000	9.77080198	5
0.00000000	12.00000000	6
1.00000000	4.00000000	7
1.17469361	5.07503763	8
1.38432595	6.36508278	9
1.63588476	7.91313696	10
1.93775532	9.77080198	11
2.30000000	12.00000000	12
2.00000000	4.00000000	13
2.40313911	5.07503763	14
2.88690604	6.36508278	15
3.46742636	7.91313696	16
4.16405074	9.77080198	17
5.00000000	12.00000000	18
3.00000000	4.00000000	19
3.50352873	4.80564597	20
4.10776320	5.77242113	21
4.83284457	6.93255132	22
5.70294222	8.32470755	23
6.74705939	9.99529503	24
8.00000000	12.00000000	25
4.00000000	4.00000000	26
4.61939141	4.61939141	27
5.36266110	5.36266110	28
6.25458474	6.25458474	29
7.32489309	7.32489309	30
8.60926312	8.60926312	31
10.15050716	10.15050716	32
12.00000000	12.00000000	33

ELEMENT NO	NODAL POINTS OF THE ELEMENT		
1	1	2	8
2	1	8	7
3	2	3	9
4	2	9	8
5	3	4	10
6	3	10	9
7	4	5	11
8	4	11	10
9	5	6	12
10	5	12	11
11	7	8	14
12	7	14	13
13	8	9	15
14	8	15	14
15	9	10	16
16	9	16	15
17	10	11	17
18	10	17	16
19	11	12	18
20	11	18	17

21	13	14	20
22	13	20	19
23	14	15	21
24	14	21	20
25	15	16	22
26	15	22	21
27	16	17	23
28	16	23	22
29	17	18	24
30	17	24	23
31	18	25	24
32	19	20	27
33	19	27	26
34	20	21	28
35	20	28	27
36	21	22	29
37	21	29	28
38	22	23	30
39	22	30	29
40	23	24	31
41	23	31	30
42	24	25	32
43	24	32	31
44	25	33	32

(1) Read/write number of lines and weighting factor.
(2) Read/write number of intervals, coordinates of starting and ending
     points of generating lines.
(3) Calculate node numbers and their corresponding x, y coordinates.
(4) Mesh variation in Fig.6.1(b).
(5) Mesh variation in Fig.6.1(a).
(6) Mesh variation in Fig.6.1(c).
(7) Skip a triangle if fourth node of quadrilateral is zero.

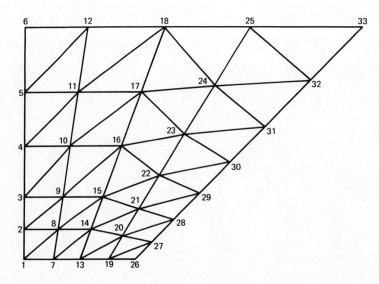

Fig.6.3 Example problem

6.2.3 ISOPARAMETRIC ELEMENTS

In this section a simple but effective mesh generation scheme using the isoparametric concept and suitable for band or front solvers will be described first. Later on a slightly more elaborate scheme suitable for front solver only will be presented to deal with mesh generation for domains with complex shapes and for multiply-connected domains.

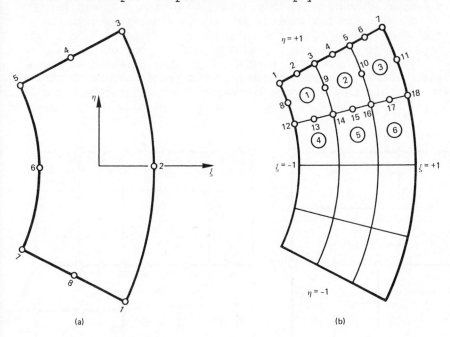

| (a) | (b) |

Fig.6.4 Curved bridge deck and example mesh

Using the curved bridge deck in Fig.6.4(a) as an example it is seen that the whole domain can be regarded as one big isoparametric element [2] and the shape is adequately defined by the eight master nodes. (Note that the circular edge is now replaced by a parabola.) A suitable mesh is then obtained by using lines of constant ξ and η to divide the 'large element' into n elements in the ξ direction and n elements in the η directions. The coordinates of any point j are given as

$$x_j = \sum_{i=1}^{8} N_i(\xi_j, \eta_j) x_i$$

$$y_j = \sum_{i=1}^{8} N_i(\xi_j, \eta_j) y_i$$

(*see* Eqns.(3.6b) and (3.6c)).

In the illustrative subroutine, the node numbering starts from $\xi = -1$, $\eta = +1$ and proceeds along the line of constant η until $\xi = +1$ is reached. For the odd numbered generating lines there are m element divisions and therefore $2m + 1$ nodes because of the presence of both corner and midside nodes. The numbering then recommences at $\xi = -1$ on the next line of constant η and the process is repeated, although this time an even numbered generating line is involved and there are only

m + 1 nodes because only midside nodes are present. All node numbers
are arrayed in ascending order as indicated in Fig.6.4(b). There is of
course no reason why the first node has to be node number 1, and in
fact the subroutine allows the user to define any number as the first
node number and then all nodes will follow sequentially. This facility
is necessary for handling domains with complex shapes.

With this simple subroutine it is possible to divide a domain into a
number of sub-domains and utilize the program to generate a mesh for
each of the sub-domains in turn, and subsequently combining them to form
a complete mesh. For example an L-shaped domain can be conveniently
divided into two 'large elements' with the ξ, η axes oriented as shown
in Fig.6.5(a). After generating the mesh for I we can repeat the process
for II, although this time the node numbering does not start at node 1,
but at node 58, which has to be input as part of the data for the second
run.

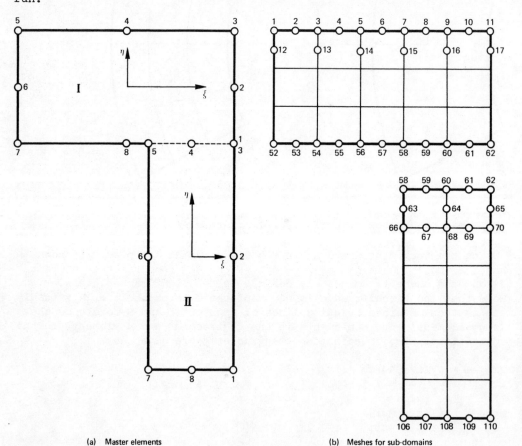

(a) Master elements   (b) Meshes for sub-domains

Fig.6.5 Mesh generation for an L-shaped domain

With a little elaboration, the same subroutine can be used to generate
a mesh for a multiply-connected domain such as a bracket area with a
hole [3] shown in Fig.6.6. As can be seen, the bracket area has been
divided into seven sub-domains.

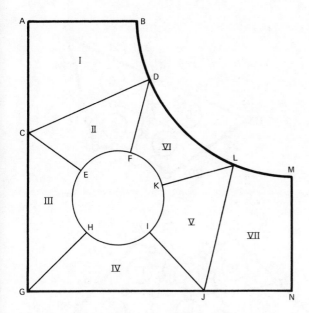

Fig.6.6 Bracket area with a hole

The subroutine is now used repeatedly to generate meshes for the seven sub-domains. The nodal numbers of the first generating line of II corresponding to the nodal numbers of the last generating line of I, and therefore no modification is necessary at the interface between I and II. For III, the first generating line does not correspond to the last generating line of II and therefore the first node number must be sequential to the last node number of II, and so on for all the other sub-domains.

In order to connect the meshes of all the sub-domains together, it is necessary to comply with the following conditions:

Along CE	17 = 38,	22 = 39,	25 = 40,	30 = 41,	33 = 42
Along GH	62 = 89,	63 = 85,	64 = 78,	65 = 74,	66 = 67
Along IJ	73 = 112,	77 = 113,	84 = 114,	88 = 115,	95 = 116
Along KL	96 = 141,	97 = 142,	98 = 143,	99 = 144,	100 = 145
Along FD	37 = 117,	32 = 118,	29 = 119,	24 = 120,	21 = 121
Along LJ	100 = 146,	103 = 151,	108 = 154,	111 = 159,	95 = 162

Note that points J and L are common to three sub-domains and therefore there are altogether three nodal numbers for each of these two points.

It is a simple matter to write several Fortran statements which will instruct the computer to replace the node numbers on the right side of the equations by the corresponding node numbers on the left side of the equations. In this way the meshes for the seven sub-domains (Fig.6.7) are connected together to form one single mesh for the multiply-connected domain.

Using the technique described above it is obvious that quite a few node numbers have been eliminated altogether. This means that a mesh generated this way is unsuitable for a band solution scheme, but is on

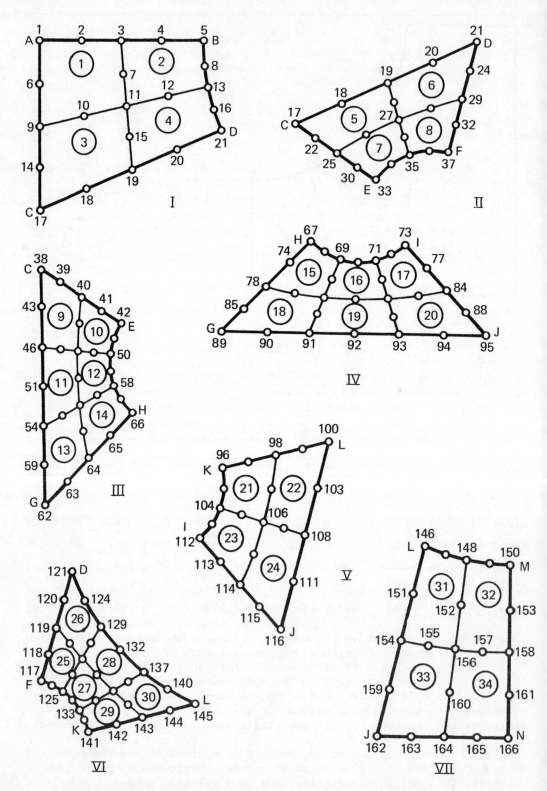

Fig.6.7 Sub-domains and their meshes

the other hand completely suitable for a front solution scheme, which is independent of node numbering.

If grading of mesh is required for the analysis, weighting factors can be introduced to create unequal divisions inside a 'large element', although a much simpler way, which is satisfactory for most problems, is to shift the origin of the ξ, η axes towards the corner in which a denser mesh is desired, as shown in Fig.6.8.

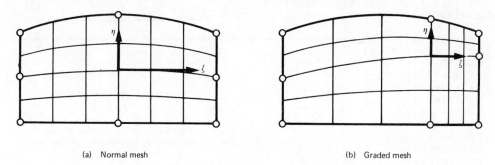

(a) Normal mesh   (b) Graded mesh

Fig.6.8 Graded mesh within a large element

Program GEN

```
      DIMENSION XM(8),YM(8),X(1000),Y(1000),LDEF(200,8),NS(200)
      DIMENSION SHP(8),XX(8),YY(8)
      DATA XX/1.0,1.0,1.0,0.0,-1.0,-1.0,-1.0,0.0/
      DATA YY/-1.0,0.0,1.0,1.0,1.0,0.0,-1.0,-1.0/
      READ(5,1000) NXL,NYL,NN,ND
      WRITE(6,1001) NXL,NYL,NN,ND
      WRITE(6,1013)
      DO 5 I=1,8
      READ(5,1002) XM(I),YM(I)
      WRITE(6,1003) I,XM(I),YM(I)
    5 CONTINUE
      II=0
      NSF1=2*NYL+1
      DO 50 I=1,NSF1
      YL=1.0-2.0*(I-1)/(2*NYL)
      II=II+1
      NS(I)=NN
      IF((II.EQ.3) II=1
      NSF2=2*NXL+1
      DO 45 J=1,NSF2,II
      XL=-1.0+2.0*(J-1)/(2*NXL)
      DO 25 IJ=1,8
      GO TO(10,15,10,20,10,15,10,20),IJ
   10 SHP(IJ)=0.25*(1.0+XL*XX(IJ))*(1.0+YL*YY(IJ))*
     1                    (XL*XX(IJ)+YL*YY(IJ)-1.0)
      GO TO 25
   15 SHP(IJ)=0.5*(1.0+XL*XX(IJ))*(1.0-YL*YL)
      GO TO 25
   20 SHP(IJ)=0.5*(1.0+YL*YY(IJ))*(1.0-XL*XL)
   25 CONTINUE
      XXX=0.0
      YYY=0.0
      DO 30 K=1,8
      XXX=XXX+SHP(K)*XM(K)
      YYY=YYY+SHP(K)*YM(K)
   30 CONTINUE
```

① ② ③ ④ ⑤ ⑥ ⑦ ⑧

```
      X(NN)=XXX
      Y(NN)=YYY
      NN=NN+1
   45 CONTINUE
   50 CONTINUE
      MAXNOD=NN-1
      NSS=NS(1)
      WRITE(6,1014)
      DO 60 NIC=NSS,MAXNOD
      WRITE(6,1020) NIC,X(NIC),Y(NIC)
   60 CONTINUE
      NEL=1
      MEL=NYL
      IF(ND.EQ.0) MEL=1
      DO 150 I=1,NYL
      IF(ND.EQ.1) NEL=I
      NSS=(I-1)*2+1
      N1=NS(NSS)
      N2=NS(NSS+1)
      N3=NS(NSS+2)
      DO 140 J=1,NXL
      LDEF(NEL,1)=N3  +(J-1)*2
      LDEF(NEL,2)=N3+1+(J-1)*2
      LDEF(NEL,3)=N3+2+(J-1)*2
      LDEF(NEL,4)=N2+1+J-1
      LDEF(NEL,5)=N1+2+(J-1)*2
      LDEF(NEL,6)=N1+1+(J-1)*2
      LDEF(NEL,7)=N1  +(J-1)*2
      LDEF(NEL,8)=N2  +J-1
      NEL=NEL+MEL
  140 CONTINUE
  150 CONTINUE
      MAXNEL=NXL*NYL
      WRITE(6,1015)
      DO 170 NEL=1,MAXNEL
      WRITE(6,1010) NEL,(LDEF(NEL,I),I=1,8)
  170 CONTINUE
      STOP
 1000 FORMAT(4I10)
 1001 FORMAT(1H1,4X,*NUMBER OF DIVISIONS IN LOCAL X DIRECTION =*,I3,
     1  //,5X,*NUMBER OF DIVISIONS IN LOCAL Y DIRECTION =*,I3,
     2  //,5X,*FIRST NODE NUMBER =*,I3,
     3  //,5X,*DIRECTION PARAMETER =*,I3)
 1002 FORMAT(2F10.3)
 1003 FORMAT(1X,I5,4F10.3)
 1010 FORMAT(10X,9I5)
 1013 FORMAT(///////,5X,*MASTER NODAL CO-ORDINATES*,//,
     1  2X,*NODE*,9X,1HX,9X,1HY)
 1014 FORMAT(///////,5X,*NODAL CO-ORDINATES*,//,
     1  7X,*NODE*,9X,1HX,9X,1HY)
 1015 FORMAT(///////,5X,*ELEMENT DEFINITIONS*)
 1020 FORMAT(6X,I5,2E10.3)
      END
```

Program output for Area II of Fig. 6.5.

```
NUMBER OF DIVISIONS IN LOCAL X DIRECTION =  2

NUMBER OF DIVISIONS IN LOCAL Y DIRECTION =  6

FIRST NODE NUMBER = 58

DIRECTION PARAMETER =  0
```

MASTER NODAL CO-ORDINATES

NODE	X	Y
1	30.000	20.000
2	20.000	20.000
3	10.000	20.000
4	10.000	16.000
5	10.000	12.000
6	20.000	12.000
7	30.000	12.000
8	30.000	16.000

NODAL CO-ORDINATES

NODE	X	Y
58	.100E+02	.120E+02
59	.100E+02	.140E+02
60	.100E+02	.160E+02
61	.100E+02	.180E+02
62	.100E+02	.200E+02
63	.117E+02	.120E+02
64	.117E+02	.160E+02
65	.117E+02	.200E+02
66	.133E+02	.120E+02
67	.133E+02	.140E+02
68	.133E+02	.160E+02
69	.133E+02	.180E+02
70	.133E+02	.200E+02
71	.150E+02	.120E+02
72	.150E+02	.160E+02
73	.150E+02	.200E+02
74	.167E+02	.120E+02
75	.167E+02	.140E+02
76	.167E+02	.160E+02
77	.167E+02	.180E+02
78	.167E+02	.200E+02
79	.183E+02	.120E+02
80	.183E+02	.160E+02
81	.183E+02	.200E+02
82	.200E+02	.120E+02
83	.200E+02	.140E+02
84	.200E+02	.160E+02
85	.200E+02	.180E+02
86	.200E+02	.200E+02
87	.217E+02	.120E+02
88	.217E+02	.160E+02
89	.217E+02	.200E+02
90	.233E+02	.120E+02
91	.233E+02	.140E+02
92	.233E+02	.160E+02
93	.233E+02	.180E+02
94	.233E+02	.200E+02
95	.250E+02	.120E+02
96	.250E+02	.160E+02
97	.250E+02	.200E+02
98	.267E+02	.120E+02
99	.267E+02	.140E+02
100	.267E+02	.160E+02
101	.267E+02	.180E+02

```
102    .267E+02    .200E+02
103    .283E+02    .120E+02
104    .283E+02    .160E+02
105    .283E+02    .200E+02
106    .300E+02    .120E+02
107    .300E+02    .140E+02
108    .300E+02    .160E+02
109    .300E+02    .180E+02
110    .300E+02    .200E+02
```

```
ELEMENT DEFINITIONS
       1    66    67    68    64    60    59    58    63
       2    68    69    70    65    62    61    60    64
       3    74    75    76    72    68    67    66    71
       4    76    77    78    73    70    69    68    72
       5    82    83    84    80    76    75    74    79
       6    84    85    86    81    78    77    76    80
       7    90    91    92    88    84    83    82    87
       8    92    93    94    89    86    85    84    88
       9    98    99   100    96    92    91    90    95
      10   100   101   102    97    94    93    92    96
      11   106   107   108   104   100    99    98   103
      12   108   109   110   105   102   101   100   104
```

(1) ξ, η coordinates of the master nodes.
(2) Read/write control data.
   NXL = number of elements in ξ direction
   NYL = number of elements in η direction
   NN = first node number
   ND = direction parameter. If ND = 1 then elements are numbered
         sequentially along lines of constant ξ. If ND = 0 the
         sequential numbering is along lines of constant η.
(3) Read/write x, y coordinates of master nodes.
(4) η coordinates of each generated node.
(5) NS( ) stores a list of the node numbers on $\xi = -1$.
(6) ξ coordinates of each generated node.
(7) Calculate the shape functions for the ξ, η coordinates of each
    generated node.
(8) Calculate the x, y coordinates of each node from

$$\begin{Bmatrix} X \\ Y \end{Bmatrix} = \sum_{i=1}^{8} N_i \begin{Bmatrix} X_{mi} \\ Y_{mi} \end{Bmatrix}$$

(9) Write out the nodal coordinates.
(10) MEL is the increment in the element numbers in ξ direction.
(11) Node numbers along edge $\xi = -1$ for use in generating the first
     element definition of a row of elements in the ξ direction.
(12) Node numbers of elements along a row of elements in the ξ direction
     using the sequential numbering of the nodes.
(13) Write out the element definitions.

References

1. W.R. Buell and B.A. Bush. Mesh generation - a Survey. *Journal of Engineering for Industry, Transactions of the American Society of Mechanical Engineers, Series B*, vol.95, no.1, Feb 1973, pp.332-338.
2. O.C. Zienkiewicz and D.V. Phillips. An automatic mesh generation scheme for plane and curved surfaces by isoparametric coordinates, *International Journal for Numerical Methods in Engineering*, vol.3, 1971, pp.519-528.
3. J. Suhara and J. Fukuda. Automatic mesh generation for finite element analysis. In *Advances in Computational Methods in Structural Mechanics and Design* (J.T. Oden, R.W. Clough and Y. Yamamoto, eds). UAH Press, Alabama, 1972, pp.607-624.

Index